Intelligent and Efficient Transport Systems - Design, Modelling, Control and Simulation

Edited by Truong Quang Dinh

Published in London, United Kingdom

IntechOpen

Supporting open minds since 2005

Intelligent and Efficient Transport Systems - Design, Modelling, Control and Simulation
http://dx.doi.org/10.5772/intechopen.81308
Edited by Truong Quang Dinh

Contributors
Yaxing Ren, Araan Mohanadass, Jj Chong, Peter J Kay, Wei-Chin Chang, Josh Sperling, Alejandro Henao, Santhosh T K, Suthanthira Vanitha N, Ramkumar K, Igor Bolvashenkov, Jörg Kammermann, Ilia Frenkel, Hans-Georg Herzog, Ahmed El-Mowafy, Nobuaki Kubo, Anh-Tu Nguyen, Thierry-Marie Guerra, Jimmy Lauber

First published in London, United Kingdom, 2020 by IntechOpen
IntechOpen is the global imprint of INTECHOPEN LIMITED, registered in England and Wales, registration number: 11086078, 7th floor, 10 Lower Thames Street, London, EC3R 6AF, United Kingdom
Printed in Croatia

British Library Cataloguing-in-Publication Data
A catalogue record for this book is available from the British Library

Additional hard and PDF copies can be obtained from orders@intechopen.com

Intelligent and Efficient Transport Systems - Design, Modelling, Control and Simulation
Edited by Truong Quang Dinh
p. cm.
Print ISBN 978-1-78984-103-9
Online ISBN 978-1-78984-104-6
eBook (PDF) ISBN 978-1-83880-511-1

An electronic version of this book is freely available, thanks to the support of libraries working with Knowledge Unlatched. KU is a collaborative initiative designed to make high quality books Open Access for the public good. More information about the initiative and links to the Open Access version can be found at www.knowledgeunlatched.org

For EU product safety concerns:
IN TECH d.o.o., Prolaz Marije Krucifikse Kozulić 3, 51000 Rijeka, Croatia, info@intechopen.com or visit our website at intechopen.com.

Meet the editor

Truong Quang Dinh (BEng, 2006; DEng, 2010) is currently an assistant professor in WMG, the University of Warwick, leading the research in energy system management and control. He has published over 150 high-quality journal and conference papers, four patents, and three books/chapters. He has delivered more than 15 keynote speeches and invited talks at renowned conferences and industrial events. He has received more than 10 journal/conference paper awards. He is a member of the Technical Committee on Motion Control of the IEEE Industrial Electronics Society, and Secretary of IEEE Computational Intelligence Society—Region 8 UKRI. He is editor and lead guest editor of six international journals, and chair and committee member of several renowned conferences.

Contents

Preface

There are urgent demands for low-carbon and energy-saving technologies across all transportation sectors. There is a need to improve energy efficiency of traditional engine-driven vehicles, which still take a large percentage of the global market and community, especially for off-road transport. In parallel, propulsion electrification is acknowledged as a promising solution to provide a system with more potential to achieve optimal energy efficiency, significant energy-saving opportunities, and low/zero emissions.

More attention has also been paid to intelligent vehicles, which help to improve system flexibility to minimize environmental pollutants, while maximizing customer satisfaction and underpinning future smart cities. However, complex structures, multidomain operations, and different transient dynamics of such propulsion systems lead to critical issues and challenges in system design, assessment, modeling, control, and management.

During the system development process, due to system safety and reliability requirements, rigorous performance evaluations need to be performed. With the rapid development of digital technologies and computer science, real-time simulation platforms are increasingly deployed as efficient design, rapid prototyping, and validation tools by both researchers and OEMs to bridge the gap between fundamental research and practical implementation/commercialization.

In this book, the discovery of digital and technology solutions and critical analyses across transportation sectors and infrastructures is introduced to address the following interesting questions:

- How can the energy efficiency of a vehicle be evaluated (Chapter 1)
- How can an electric propulsion system be designed and evaluated (Chapter 2)
- What are the potential opportunities for intelligent transportation systems and smart cities? (Chapters 3–4)
- How can transportation systems and infrastructure performance be improved through control and energy-power management? (Chapters 5–7)
- What are the potential modeling tools and real-time simulation platforms available? (Chapters 5–9)

It is exciting to know that the contents presented in this book are the work of practitioners, researchers, scientists, and scholars from different countries, including the United Kingdom, France, the United States, Japan, India, Taiwan, Ethiopia,

Israel, and Australia. This book will be useful to a wide range of audiences: university staff and students, engineers, and business people working in relevant fields.

Dr. Truong Quang Dinh
Energy System Management and Control,
International Automotive Research Centre (IARC), WMG,
University of Warwick,
Coventry, United Kingdom

Chapter 1

Making the Most of the Energy We Have: Vehicle Efficiency

Araan Mohanadass

Abstract

With global concerns over emissions from non-renewable sources and its dwindling global supplies. Optimization of our energy usage is highly important. Converting energy to various forms is usually an imperfect process with energy being wasted. Vehicle's convert on-board stored energy to a kinetic form to drive a vehicle. Understanding the sources of energy losses allow us to (a) reduce emissions created by combustion engines, and (b) improve driving range of electrified powertrains. Within this chapter energy losses across the vehicle will be explored for both non-powertrain and powertrain components. Modes of losses and test methods for vehicle components will also be outlined to highlight their significance in vehicle efficiency.

Keywords: vehicle efficiency, non-powertrain, powertrain, drag, combustion engines, electrification

1. Introduction

Energy is finite, as stated by the rules that govern the physical universe. It can neither be created nor destroyed, only stored in bonds between sub-atomic and atomic particles. Only when bonds are established or are broken is energy released. This occurs until a new near stable state is achieved. This is a universal rule, which has yet to be modeled any better.

Until the turn of the twenty-first century, fossil fuels have been the most practical way to power vehicles and most utilities in the world. These fuels were abundant in large quantities, could be transported with ease to all corners of the globe, the control of the exothermic was well understood and was considered a lot safer than other sources of power. For majority of the century it was the ideal energy carrier, but this was short sighted for two reasons. Firstly, the oil wells that are used are a finite supply. The vast fossil fuel reserves that we have been draining from were created by millennia of decomposed biomass that was present almost 650 million years ago. As current non-renewable supplies are being consumed at a faster rate, and the likelihood of finding new fossil fuel sources decrease. There will likely be a point where the once ideal energy carrier will be completely consumed. From recent studies, this is expected to occur in 50 years for crude oil and natural gas, while half a century more for coal [1, 2]. As we continue to consume this finite supply, the value of the fuel will increase with a large impact to all economic sectors. The only way to avoid this is by reducing our energy consumption or looking for alternatives energy sources.

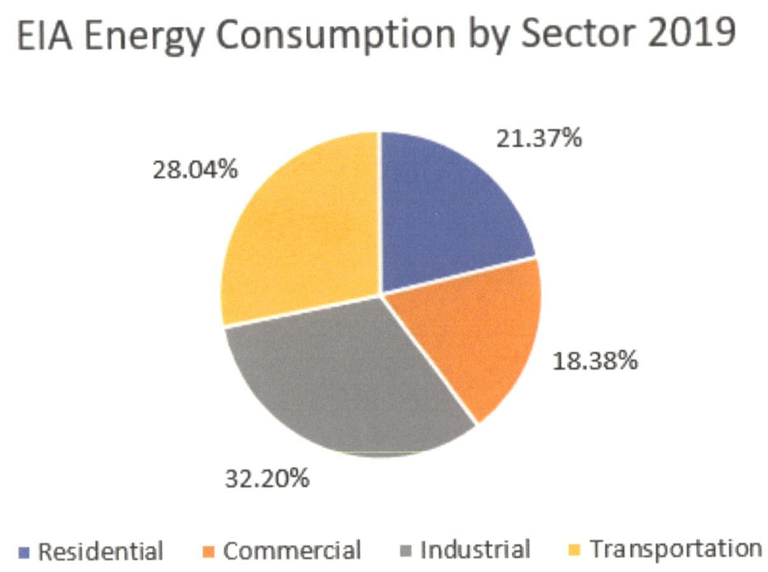

Figure 1.
The total US energy consumption by sector in 2019.

The transportation sector last year alone consumed an estimated 115.5 quadrillion Btu of energy [3]. As a prominent global energy consumer, improvements to transport energy efficiency can make a significant impact on a global scales (**Figure 1**). To act on this transport engineers have focused on improving current combustion technology, deployment of powertrain electrification and improvement to non-powertrain components.

2. The fundamentals of energy consumption for vehicles

Vehicle efficiency is generally defined by the distance traveled per use of a quantified energy. Most commonly, for the last century the vehicle metric for efficiency has been Mile per gallon of fuel (MPG). This metric was used universally among most countries due to legislation sharing. With refinements to emission impact measurements and alternate vehicle fuel source, new metrics such as CO_2 grams per kilometer and kilowatt-hours per kilometer have been adopted. With the introduction of electrification into vehicle power-trains the latter unit is used to quantify vehicle efficiency.

Breaking down to the basics of vehicle efficiency, we can explain the energy needed to move a vehicle using Newtonian mechanics. Using Eq. (1), we can deduce that variation in vehicle force can accelerate and decelerate the vehicle. A positive force on the left-hand-side (LHS) of the Eq. (1) means that the vehicle will be accelerating, a negative value means the vehicle is decelerating while a zero net force means the vehicle is in an equilibrium state (not moving or a constant velocity). The value of this force itself is a summation of the driving force provided by the vehicle's power-plant and the drag force that resists the vehicles motion. Vehicle force can be represented as power, which in turn can be used to calculate the energy required to move the vehicle.

$$Forces = Vehicle\ \ Mass \times Acceleration \tag{1}$$

In addition to the force produced by the vehicle, the vehicle mass has an influence towards how the transformation of energy is directed to the vehicle's motion. This is known as inertial mass. As an example, consider pushing a shopping trolley at the start of trip to the supermarket. When the trolley is empty it is easier push and as a result easier to accelerate, we can consider the trolley as having low inertia. In comparison at the end of the shopping trip when the trolley is full and has a higher mass it requires more force to reach the same acceleration and therefore described as having a higher inertia. The same principle occurs with vehicles, the heavier they are the harder it is for them to accelerate/decelerate. To overcome this requires either more power from the vehicle or a reduction vehicle inertia. Reduction in vehicle inertia has been a popular way to improve vehicle efficiency over the last several decades. The adaptation of light-weighting can be seen in the changes to chassis design and incorporation of new materials.

3. Forces acting on a vehicle

As previously mentioned a vehicle accelerates and decelerates by producing a net force. This net driving force is a combination of a propulsion force produced by the powertrain, and the drag forces that resists the vehicles motion Eq. (2). Both types of forces consume energy with the drag forces being parasitic in nature.

$$Driving\ \ Forces = Propulsion\ \ Force - Drag\ \ Force \tag{2}$$

3.1 Propulsion force

The vehicle powertrain provides the propulsion force needed to put the vehicle in motion as intended by the driver. Vehicle powertrains create this force by converting stored energy on-board to usable kinetic power providing force/torque. Typically with modern vehicles, hydro-carbon fuels or electro-chemical cells are used as the energy storage medium.

3.2 Vehicle drag forces

Vehicle drag is formed of numerous contributors across the vehicle. The nature of parasitic losses are frictional, viscous and inertial sourced. The absorbed energy is dissipated in impractical power that cannot be used to propel the vehicle. In this section, the losses across the tire, drive-train transmission and aerodynamic drag will be explored.

3.2.1 Tire drag

The wheels of a vehicle allow the powertrain to apply the driving force to the road surface. As the wheel has torque applied to it, a force output is transferred to the contact patch where the tire meets the road. This action can be represented by Eq. (3). The design of a wheel however incorporates the tire. The pneumatic tire commonly used on road vehicles are designed to not only transfer the driving force but also absorb shocks across the road. Because of this not all power is passed onto the road.

$$Tire\ \ Drag = \frac{Torque}{Tire\ \ Radius} \tag{3}$$

The primary reason for this energy loss is due to the deformation of the tire. It is mistakenly believed that the tire itself is perfectly circular when in contact with the road. With the overall weight of the vehicle acting on the wheel, the contact patch of the wheel is flattened disrupting the wheel's circular shape. As the tire rotates the contact patch moves across the wheel in relation to the ground. This motion requires the tread and sidewall directly in front of the contact patch to compress, then expand across the contact patch until it finally relaxes past the road surface [4, 5]. The tires rubber has elastic material properties and as it is compressed or expanded work is done, therefore consuming power provided by the driving force of the vehicle.

The drag produced by the tire is normally considered a function of the tires geometry, vertical load, pneumatic pressure and velocity. Furthermore, co-efficient can be established to capture influences from the tires construction, road surface and suspension settings. The drag force can be represented in a vehicle model by either maps or equations. A widely used tire drag model is the SAE J2452 equation Eq. (4) [6]. This equation utilizes five co-efficient that work in conjunction with the tire load, pneumatic pressure and velocity to provide the tire drag. The co-efficient for this model are achieved through physical testing.

$$Tire\ \ Drag = Load^{\alpha} \cdot Tire\ \ Pressure^{\beta} \cdot \left(A + BV + CV^2\right) \tag{4}$$

Component level drag testing for tires can be done in numerous ways. Tires are usually attached to a instrumented axle and then spun up to vehicle speeds while on a simulated road surface. The simulated road surface are the differentiating factor between most tire drag tests. Drum rigs where the tire sits upon a rotating drum and belt rigs that recreate the road using a conveyor style belt are the most common techniques for tire drag testing. Both provide highly controlled environments allowing for precise testing. However, the inclusion of simulated roads lead to some deviation in accuracy. Most visibly, the overestimation of tire drag through the drum method. This is due to the curvature of the drum imposed on the wheel, leading to a higher vertical deflection at the contact patch [4]. This deviation from the real life road conditions is the most significant draw back to most tire testing methods. More recently sophisticated mobile test tire trailer (MTTT) have been utilized to overcome the accuracy issues with simulated roads. This method of testing similarly uses an instrumented axle, but attaches it underneath an articulated trucks trailer. In this configuration the tire can be tested accurately to that of a real world case [7]. Though this method is the most accurate test method, it does sacrifice precision due to the reduced control of the test environment.

3.2.2 Drivetrain drag

All powertrains have regions in their operating windows where they provide power most effectively. Outside of these thresholds the powertrain unit can become inefficient. Typically this is dependent on the powertrains speed of operation. To overcome this, engineers and designers utilize gears and mechanical transmission devices to operate powertrain in their efficiency bands. The drivetrain's gearbox, differential and power-split devices share similar efficiency characteristics.

Drivetrain units suffer from two primary loss modes. The design of drivetrain components incorporate a multitude of gears, bearings and clutches to convert power. From the gears and bearings, natural imperfection at the individual component level can lead to friction that consumes the driving power. The lost power through friction is normally released as heat. In addition to this loss mode, transmission units also utilize oil to lubricate, cool and reduce wear around moving components. Because of this, viscous drag is experienced as the oil in the drivetrain units resist the motion of moving components [8]. Both of these drag modes are subject to the velocity and torque of the moving components. The higher both of the factors are, the higher the drag will be. Moreover, oil temperature will also have an influence on the viscous drag forces. As with most liquids including transmission oil, an increase in oil temperature leads to an increase of oil viscosity. Therefore at higher temperatures the viscous drag is reduced across the drive-train units. It is however not wise to operate drive-trains at high temperatures as it is to operate them at cold temperature. At high temperatures, design tolerances may be affected as parts expand and become weaker. While at cold temperatures viscous drag forces increase.

The calculation and estimation of drag losses are typically achieved either by computer-aided design/engineering (CAD/CAE) software or through physical testing [9]. Typical gearboxes, differential and power-split devices can include several bearings, clutches and various internal geometries that interact with lubricant flow making energy equations difficult to manage. With CAD/CAE tools, drive-train component drag can be estimated with the design and material choices known. This modeling technique for energy efficiency provides an adequate energy benchmark for rough assumptions. Though the actual component can be mocked up precisely in CAD/CAE, viscous and thermal interaction can be difficult to model accurately in simulations. Consequently, CAD/CAE efficiency results are limited to initial vehicle efficiency analysis before prototypes are built. On the other hand, physical testing can produce accurate energy characteristics of drivetrain components once prototype or production has been established. With the physical components, general energy consumption can be observed through analysis of power input and output. Dynamometers connected to the input and output shafts of the drive train unit are used to obtain these results. Steady-state tests are usually run with various set-points to obtain measurements of both input shaft velocity and torque inputs. The effect of temperature can also be included in component tests by either letting the component naturally soak thermal energy from a temperature controlled environment or by artificially controlling the oil temperature to desired levels. Results are constructed into tabulated maps indicating lost torque, lost energy or efficiency percentage.

3.2.3 Aerodynamic and hydrodynamic drag

Aerodynamic and hydrodynamic drag acts on the body of a vehicle moving through a volume of gas or liquid. Both aerodynamics and hydrodynamics share the same principles and are only separated by the medium that they are being propelled through. For the rest of this section we will look at the effects of drag on ground based vehicles, but the fundamental principles are directly transferable to marine vehicles.

Aerodynamic drag exists while a volume of air passes over the vehicle. As the air passes, it opposes the vehicle through various viscous drag dynamics. Over the last several decades energy efficiency optimization through aerodynamic drag reduction has been a noticeable through vehicle design. Majority of passenger road vehicles pre-1990 had a box-styled shape. As the understanding of aerodynamics grew and capability in vehicle body manufacturing expanded, passenger vehicles became sleeker. Not just for their improved esthetics, but also to utilize existing driving power without up-sizing power-plants. Today across all vehicle streams, manufacturers and designers try to exploit aerodynamics to reduce drag of vehicles using various tools.

Aerodynamic drag is a culmination of two primary interactions between the vehicle body and the fluid volume. To begin with a portion of fluid drag is produced across the surface of the vehicle body. This is known as skin friction and can be attributed to interactions of the fluid at the vehicles surface. Air is a viscous fluid, and behaves similar to that of its liquid counterparts which have lower viscosity. Viscosity of a fluid denotes the strength of the inter-molecular attraction between molecules in a fluid. As a fluid goes through a laminar flow over a surface, the immediate molecules at the surface can be assumed to have zero velocity. These molecules through their viscous properties exhibit a shear stress on the next layer of fluid molecules slowing them down from the free velocity stream. This occurrence of slowing down the next molecule layer continues until the next layer reaches the velocity of the free air-stream (**Figure 2**). This interaction is known as the boundary layer. With larger viscous forces acting between fluid molecules the distance at which the fluid layer return to free stream velocity grows. The increased size of this boundary layer contributes to a higher skin friction. The factors of control for designers and engineers on skin friction are limited. It is impractical and almost impossible to change the fluid's viscous properties. The only remaining factors that can alter skin friction are the body's overall surface area and the type of airflow over a surface. In the past aerodynamicists have experimented with the use of turbulent flow at the body's surface as an attempt to reduce skin friction drag. By introducing turbulence at the body's surface, boundary layers are disrupted therefore reducing boundary layer skin friction. The adoption of using turbulent flow to modify skin friction has however fallen out of favor due to the complexity of incorporating turbulence generating features in vehicle design and the additional pressure drag created by turbulence.

In addition to skin friction, the volume of the body and its interaction of displacing the fluid medium creates pressure drag. As the body displaces fluid, a volume behind the vehicle is left with low pressure as the air-stream attempts to return to an equilibrium state. With reduced pressure behind the vehicle, a pressure difference is created which results in drag. The key contributors for pressure drag is the body's shape, air stream velocity and the fluid properties. A large portion of aerodynamic drag is contributed by pressure drag, the typical aerodynamic drag equation is based around its properties. The first of which is the impact from of vehicle's co-efficient of drag (see Eq. (5)). As described previously the fundamental reason for pressure drag is the reduced volume of pressure created behind the

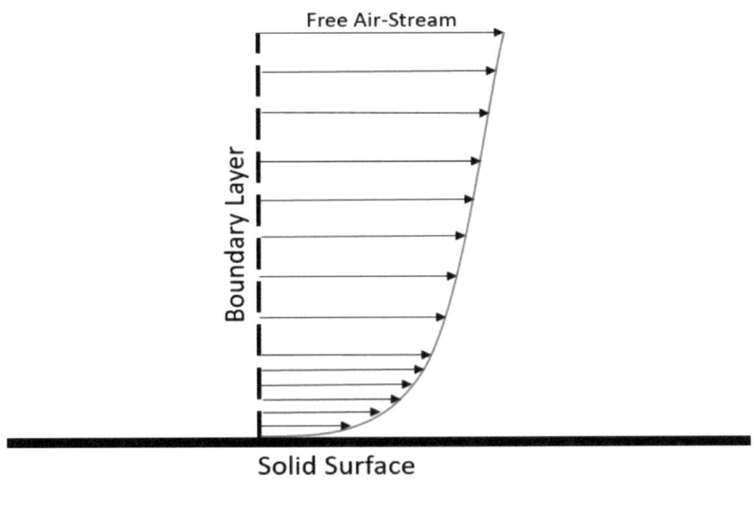

Figure 2.
Skin friction experienced at a vehicle's surface due to boundary layer interactions.

vehicle due to the displacement of the fluid medium. Vehicle designer can streamline the form of the vehicle to aid with the fluid returning to an equilibrium state behind the vehicle reducing the pressure differential on the vehicle. The streamline characteristic of the vehicle can be quantified in the co-efficient of drag metric and the frontal surface area of the vehicle.

The principle of aerodynamics emphasize the importance of drag reduction for high speed laminar flow vehicles. Vehicles at operating higher speeds will undergo a vast increase in aerodynamic drag. As seen with the aerodynamic equation, the velocity of the free-stream fluid increases drag at an exponential rate (see Eq. (5)). This characteristic of aerodynamic drag is a key focus for vehicle designers.

To quantify a vehicle's drag characteristics the co-efficient of drag and frontal surface area of a vehicle must be quantified. The frontal area can be found through physical measurements or design calculations. The co-efficient of drag however can only be quantified through physical testing or computational fluid dynamics (CFD). Through physical testing, wind/water tunnels and basins are used with models or actual vehicles to obtain this metric. The accuracy of these physical tests depend on the replication of environmental conditions. An example of such, is replication of moving roads and spinning wheels for on-road vehicles. Neglecting these factors could lead to inaccuracies in the co-efficient of drag measurement of the complete vehicle. Some vehicle specific aerodynamic test facilities have incorporated these features overcoming this potential issue, but with inclusion of such features, wind tunnel testing remains a costly option.

$$Aerodynamic\ \ Drag = \frac{1}{2} Cd\ \ Frontal\ \ Area\ \ \rho V^2 \tag{5}$$

As an alternative, aerodynamic testing can be achieved through computational means, using mathematical models based on Navier–Stokes and other models to represent turbulent interactions. Vehicle models are broken down to create a finite volume mesh and once that is achieved individual flow equations can be determined per cell. The accuracy of CFD simulations rely on the replication correct fluid dynamics and the resolution of the mesh created. With complex shapes, refined meshes are needed. With an increased number of mesh cells, an increased number of computations are needed and sufficient additional computation time to store and re-call the meshed data from memory. An example of this can be seen in **Figure 3**. As with this model of a light-weight race car, an increased mesh resolution is needed around the front wing where significant flow activity will occur. To overcome this, high resolution CFD simulations are run on supercomputers to reduce overall computational time.

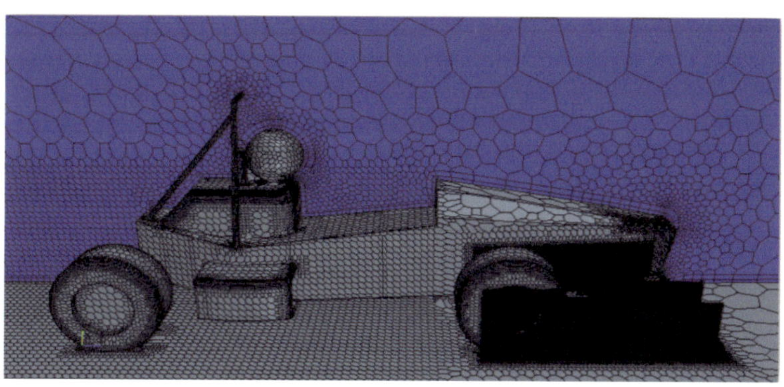

Figure 3.
A diagram displaying meshing resolution of a ultra-light-weight race vehicle.

Figure 4.
The visualization of air stream flow over a front wing without an end-plate foot (left) and a front wing with an end-plate foot (right).

CFD has become a very popular technique due to its low costs and benefits in qualitative analysis. Tests can be repeated with little costs, allowing iterative design processes to be sped up. An example of such can be seen in **Figure 4** with the iterative design process used to implement end-plate foots on a race vehicle's front wing. In addition, various visualization techniques can be used to make improved analysis towards understanding fluid flow.

3.2.4 Summary of parasitic losses

The parasitic drag modes discussed can all be tested and represented in a single equation. As outline by the World Light vehicle Test Procedure (WLTP), a vehicles parasitic drag contribution can be tested during vehicle coast down tests. Through this procedure a vehicle is allowed to decelerate under its own drag. By measuring its deceleration at certain velocity points the drag force can be interpreted by Eq. (6). The formation of the equation captures the mechanical losses of the tires and drivetrain, and the aerodynamic losses. The aerodynamic portion of the equation differs from the typical aerodynamic equation as it incorporates the ability to measure the influence of crosswinds through the use of an anemometer.

$$Parasitic\ \ Drag = A_m + B_m v + C_m v^2 + \left(\frac{1}{2}\right)\rho A V_r^2 (a_0 + a_1 Y + a_2 Y^2 + a_3 Y^3 + a_4 Y^4 \quad (6)$$

4. Energy losses across the powertrain

4.1 Combustion engines

Combustion engines have been the dominant power generation device for vehicles in the twenty-first century. Using the principles of the Otto-cycle, hydrocarbon fuel in the form of gasoline and diesel undergo combustion resulting in an exothermic reaction. The released energy is used to move a piston to provide kinetic power through a crankshaft. The popularity of this powertrain grew due to the low costs of gasoline and diesel but also thanks to the greater range in comparison to the electric vehicle (EV) powertrains and quickness in re-fueling vs. re-charging.

One of the short falls for combustion engines are their overall efficiency. These type of engines generally have an efficiency of 30–40%. The Otto-cycle relies on the expansion of gases through volume and pressure to create a driving force. The reaction itself suffers from inefficiencies, the energy released during combustion also gets released in the form of heat, light and noise. As a general rule of thumb a third of the energy is used for motive force, a third lost as heat and a another third through exhaust gases.

Moreover, energy is also lost to frictional, viscous and pumping losses. As an engine rotates, its motion is opposed by friction at the crankshaft bearings, piston rod connections and pistons sleeves; viscous interaction between the crankshaft and the oil in the lower engine block; and finally the pumping losses experienced in the combustion chamber as air needs to be sucked in or exhaust gases need to be expelled out [10]. Parasitic losses are also generated by most vehicles needing alternators and pumps to runs auxiliary systems.

Estimations around powertrain efficiency can be found from calculations based of design characteristics. Using assumptions of near perfect combustion and engine characteristics such as combustion chamber volume, and number of cylinders, power output of the engine can be estimated. With the abundance of test data from previous engines, these estimations can be refined to incorporate factors such as induction methods, engine layout and combustion timings. For absolute values of engine efficiency, physical testing of prototype or production engines are carried out. This is typically achieved through engine dynamometer testing. To measure efficiency, the energy of the fuel used is compared to the output power produced by the engine.

Efficiency characteristics of Otto-cycle engines are uniform across all variations. The output of steady state tests outline the peak torque across the engine speed and a gradient map outlining efficiency. Combustion engines in most cases have a single point of peak efficiency (OOP), this is normally located midway of the engine's capable speed and found close to wide-open throttle. The losses at certain portions of the maps are due to an increase in specific loss mode. At low engine speeds friction losses attribute to the main loss mode, while at higher speeds viscous drag becomes the overriding loss. This is illustrated by **Figure 5**. Following this, engine

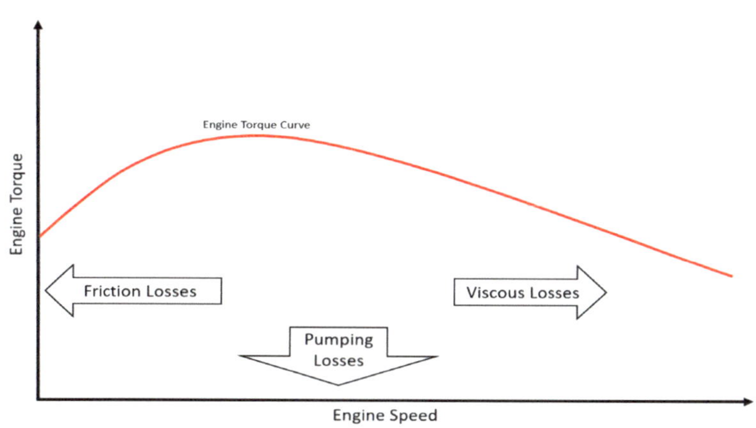

Figure 5.
Dominant losses across a combustion engines torque map.

efficiency only increases as it approaches wide open throttle due to a reduction in pumping losses resulted by a reduction in throttle restriction.

Diesel compression engines account for a large proportion of consumer and industrial vehicles. Like gasoline, diesel is made up of hydrocarbons but with a longer individual chain length. This alters the chemical properties, reducing volatility and increasing the ignition point of the fuel. As such ignition in diesel engines are achieved by auto-ignition through compression of the air-diesel fuel combo. Through high pressure fuel injection, and high compression ratios in the cylinder, combustion occurs more effectively due to the better air-fuel mixing. This is evident through the efficiency tests and reported MPG measurements from consumers. However, due to the combustion characteristics of diesel compression engines, additional emissions such as NOx and carbon particulates are increased. This is due to the leaner air-fuel mix creating a hotter combustion. Such emissions are harmful carcinogen to living creatures. Technologies to reduce these harmful emissions have been adopted, but with their inclusion the economic feasibility of diesel engines reduces.

4.2 Electric powertrains

Adoption of electrification in vehicle powertrains have grown in the last couple decades with the aspiration for energy efficient powertrains. In overall, on-road electric powertrains demonstrate energy efficiency measurements of 90%. This vast increase in efficiency is strong argument alongside with the zero immediate emissions for adoption of electrified powertrains [11].

Typically, electric powertrains have a broader range of efficiency across its operating window. As a general characteristic, EV powertrains produce large values of torque at lower speeds. At these lower speeds, electric powertrains are efficient due to reduced parasitic forces of the e-machine and low switching losses at the inverter. For vehicles that require transient power, especially at low speeds, electric powertrains are best suited.

All electrified powertrains utilize an e-machine to convert electrical energy to that of kinetic energy using magnets and electro-magnets. The control and supply of power is delivered through an alternating current. Energy storage for EVs are provided by chemical batteries. In today's adaptation of EVs, these cells usually consist of lithium electrolyte mixed with additives to improve discharge and health characteristics. By contrast, batteries typically supply only direct current, making the two components incompatible through direct connection. To overcome this, a switching device known as an inverter is used to convert the direct current (DC) of the battery to alternating current (AC) needed by the e-machine. Not only does the inverter facilitate current conversion, it also has the ability to control either the torque or speed of the e-machine. Each component of the electric powertrain suffer from individual loss modes.

4.2.1 Batteries

Though batteries are an energy storage device, an efficiency value can be attributed to ability to deliver the energy that was initially supplied to it. The primary loss method for batteries is through cell internal resistance. The cells internal resistance is a combination of the cathode, anode and electrolyte resistances [12]. As with the resistance losses of the other electrical components, a battery's power loss can be calculated through Eq. (7). The internal resistance of the battery cell is subjected to the cells temperature and its State of Charge (SOC). At cold temperatures and low SOC values the internal resistance rises rapidly (**Figure 6**). As a result to optimize

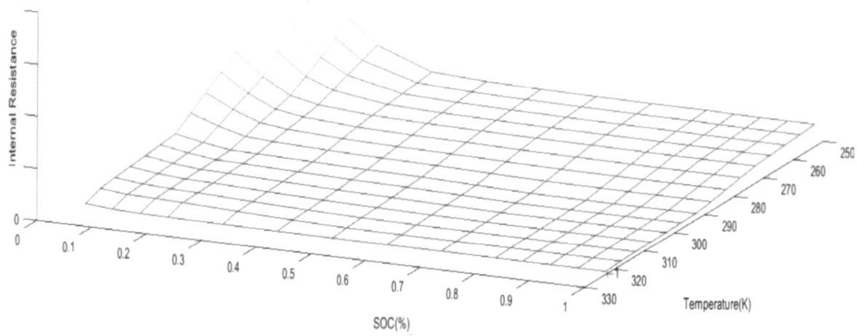

Figure 6.
The above map outlines the behavior of the internal resistance of the battery in relation to battery SOC and temperature.

efficiency, it is desirable to operate them at warmer temperature and greater SOCs. The energy lost through resistance losses are released in the form of thermal power.

$$Cell\ \ Resistance\ \ Losses\ \ (Watts) = I^2R_{anode} + I^2R_{cathode} + I^2R_{electrolyte} \tag{7}$$

The characterization of this power loss parameter is usually obtained through cell testing. Cell internal resistance can be found through either DC ohmic methods, AC pulse conduction or electro-chemical impedance spectroscopy. The specific method of choice is dependent on intended use of the battery cell. For vehicle applications, loads that are experienced by the battery are typically reactive and as a result AC pulse conduction methods are most suited. The AC method focuses on the impedance of the battery that includes not just the resistance but also the capacitance and inductance of the cell too.

A battery's capacity to store charge deteriorates over its lifespan. The measurement of available charge storage is quantified by State of Health (SOH). As a cell is charged and discharged, the quantity of useful lithium-ion electrolyte is diminished, due to decomposition of the electrolyte to Solid Electrolyte Interphase (SEI). The effect of these mechanism are outlined in **Table 1**.

The rate at which the electrolyte is lost is influenced by the depth of discharge, temperature and number of cycles experienced by the battery. Operating batteries at low SOCs, higher temperatures and large number of overall cycles will increase the rate at which the battery loses its capacity to store charge [14]. For automotive vehicles that rely on electrified powertrains, loss of stored energy capacity reduces the overall range that the vehicle can achieve on a fully charged battery. A conscious effort must be made by vehicle engineers to ensure a battery operates within a temperature window that neither degrades capacity or increases power loss (**Figure 7**).

Theoretical determination of the battery's SOH is extremely difficult. This is due to the difficulties of determining long-term physical models of internal chemical interactions. Models of battery SOH are usually achieved through experimentation. Data gathered from experiments can be utilized through regression analysis, n-term interpolation and statistical methods. The validity, computational time and effectiveness of data driven models can vary across the broad design characteristics of electro-chemical cell batteries [15].

Batteries are typically modeled theoretically in an equivalent circuit model. A battery in this model can be represented by the combination of resistors and capacitors in series and parallel. The individual resistance and capacitance values must be

Loss mechanism	Effect	Causes
Electrolyte decomposition	Power loss & capacity loss	High SoC & high temperatures
SEI growth on anode	Power loss	High SoC & high Temperatures
Contact loss of active material	Capacity loss	High cycling rate & high DoD
Current collector corrosion	Power loss	Over-discharge & low SoC
Lithium plating of the anode	Capacity loss	Low temperature, high cycling rates and poor cell balancing

Table 1.
Loss mechanisms and their effect in relation to electrolyte and anode interaction [13].

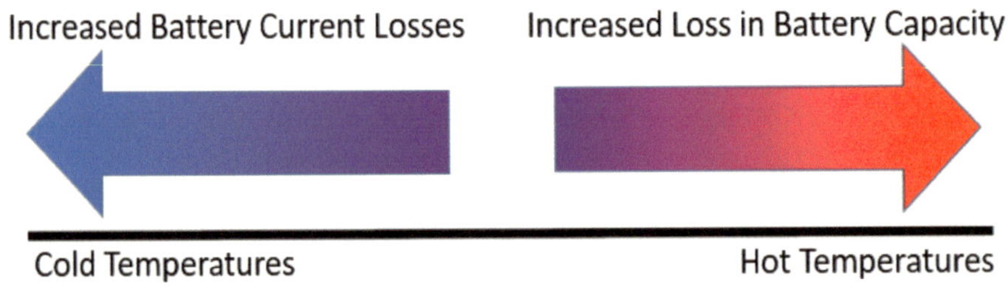

Figure 7.
Li-ion battery operation should avoid cold temperature where internal resistance losses increase or high temperatures where SOH degradation occurs.

derived from physical test data. From this configuration of the circuit, the SOC of the battery can be interpolated from the open-circuit-voltage (**Figure 8**) [16]. The model in this form can only provide representation of the battery's SOC. Any estimation of the battery's health would require adaptation of this basic model to incorporate the experimental SOH results to modify the components appropriately.

4.2.2 E-machines

The e-machines used for EVs suffer from electrical and mechanical losses. On the electrical aspects of an e-machine, power used to drive the device is lost through

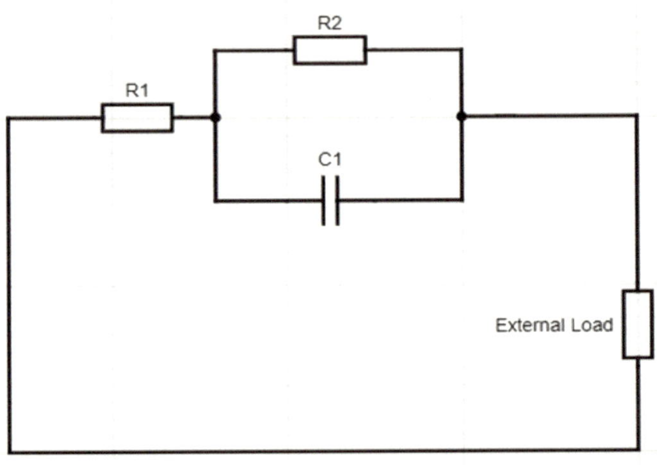

Figure 8.
An equivalent circuit model of a typical battery.

copper losses and eddy currents. Copper losses occur naturally in any current supplying conductor. Though conductors are chosen for their high conductance, they still suffer from some resistance. As current passes through, it experiences these internal resistances and power is lost. The power lost to copper losses relates to the current demanded, the losses itself can be calculated by Eq. (7).

$$Lost\ \ Power\ \ (Watts) = Current\ \ (Amps)^2 \times Resistance\ \ (\Omega) \tag{8}$$

Outside of current losses, eddy losses occur within the cores of the rotor and stator. As magnetic flux occurs across the stator and rotor, current is induced into the iron core conductors. As this is done, current loops within the core also produce their own countering magnetic field to the original as per Lenz' law [17]. As a product of the opposing magnetic field, some of the kinetic energy produced for the driving force is resisted and dissipated as heat. Mechanical losses for e-machine are very similar to losses for drivetrain components. Both viscous losses through the rotor drag (with air or coolant) and mechanical losses at the bearing will also reduce efficiency of the e-machine. The valuation of these losses can either be preliminary attained through CAD/CAE software or through physical coast-down tests. The significance of each loss varies across the operating window of the e-machine (**Figure 9**). Full characterization of motor efficiency are normally carried out through e-machine tests on dynamometers. Input power is calculated through the measurement of the AC waveform going into the motor and compared to output power derived from the torque and speed of the e-machine. In addition to powered tests, non-powered tests will also be done to breakdown mechanical losses. The combined results of which are usually represented by the use of efficiency maps.

4.2.3 Inverters

Inverters uses a H-bridge topology to switch the DC power provided by the battery to that of an AC waveform needed for the e-machine. Switching is achieved using transistors. In most cases these are an insulated-gate bipolar transistor (IGBT) or metal-oxide-semiconductor field-effect transistor (MOSFET). These type of transistors allow fast high power switching, that are controllable through low voltage signals, while operating at high efficiency values. Inverters lose energy most commonly from conduction, switching, and off state blocking losses, and are usually released as thermal energy at the transistor. Conduction losses are brought

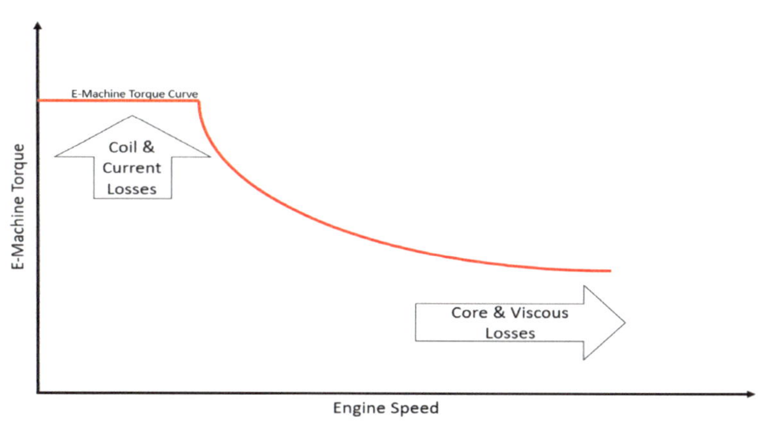

Figure 9.
Dominant losses across an E-machine torque map.

about due to forward saturation voltages and can be calculated in conjunction with the current passing through the inverter. Switching losses occur when the transistor switches states (on and off). During turn-on events capacitance between the gate and drain of transistor leads to an overlap of drain voltage and current resulting in lost energy. Turn-off energy loss behavior occurs after the gate voltage drops, during which the drain current naturally maintains until the drain voltage returns to peak. After which, the drain current logarithmically decays [18]. This overlap, as similar to that of the turn-on losses result in energy being lost. Switching losses are difficult to determine numerically and as a result loss characteristics are determined from test results. The balance of losses between modes are also a key design consideration for transistor choice. Solutions to overcome conductance losses may in fact increase switching losses. By increasing transistor size, conductance losses may be reduced, but consequently gate-drain capacitance would increase producing larger switching losses [19]. A balance must be met between the switching loss which dominates losses at high frequencies, in comparison to conductive loss dominance at low frequencies. In addition, transistors like most electrical components suffer from thermal influences. At increased temperatures, reduced efficiency can be expected from transistors. Thermal management of the inverter is important to ensure efficiency is maintained.

5. Improving energy efficiency

By understanding the mechanism from which energy is lost within automotive vehicles, we can identify trends for potential technology optimization to improve vehicle efficiency. Certain areas are optimization offer feasible opportunities that achieve the goals of regulators to reduce emission while keeping vehicles economical for consumer purchase and use.

5.1 Hybridization

Hybridization is currently a popular tool to improve vehicle efficiency. This is typically seen with a combination of a combustion engine and electric powertrain. By a combination of the two, emission produced by combustion engine can be reduced, energy management of the greater powertrain can be achieved and disturbance to current driver behavior can be minimized. Various configuration exist that vary depth of overall electrification and driving modes.

5.2 Recycling energy in combustion engines

Though the physics of combustion engines are inherent to their inefficiency, improved control of combustion characteristics and recycling lost energy can improve their efficiency. Improved design and simulation of intake and exhaust manifolds are common features on modern combustion vehicles [20]. The use of these tools help reduce the pumping losses, directly improving the cycle's efficiency.

In addition, the stability of exhaust gases can be used to alter combustion characteristics to improve efficiency. Exhaust gas re-circulation devices are now a standard feature on most diesel combustion engines. By diluting small amounts of the exhaust gas into the air intake post-throttle, allow engines to operate at wider throttle restrictions to reduce pumping losses but also reduce the overall temperature of the combustion in the engine [21]. The reduced combustion temperature aids in reducing the generation of NOx [22].

A popular method of increasing efficiency is through the use of charging the engine through turbochargers. This allows for downsizing of engines. Turbochargers use the energy of the exhaust gas to power an impeller that boost the air intake pressure. By increasing the intake pressure, the pumping losses can be further reduced as well as an increased amount of fuel-air in-cylinder leading to more power [23]. Alternatively, superchargers can increase intake pressures, but due to their reliance on the engine torque as their prime mover, increased parasitic losses are experienced. This reduces their overall impact on powertrain efficiency.

Moreover, some manufacturers have adopted selective cylinder deactivation as a method to reduce energy consumption. When the vehicle is operated in low power situations (low torque and low speed), only the necessary number of cylinders will receive fuel via their respective fuel injectors. An example of this can be seen by deactivation of four cylinders of a V8 engine developed by Daimler-Chrysler [24]. Using this approach, fuel is only needed for four cylinder over in normal situation where eight are typically used. Further, as an increased load is placed on the four active cylinders, the engine's throttle is opened further, thus reducing pumping losses [25].

5.3 Improvements in battery technology

Improvements in electric powertrain component efficiency are highly sort after. With a lower energy consumption, batteries on an electric vehicles can be reduced in size. Batteries size has a significant impact on the overall vehicle. Foremost, due to the relative recent advancement in battery technologies and their production methods, li-ion batteries seen in most electrified powertrain influence a large proportion of overall vehicle cost. In addition, with the reduced number of batteries the overall weight of the battery pack can be diminished. Majority of these advancements will come in the form of material choices for the electrolyte, cathode and anode of the battery cell. Most notably, innovative new solutions along solid-state type electrolytes have taken traction. Solid-state electrolytes provide opportunities for better power and capacity characteristics [26].

5.4 Next generation power inverter

Current inverter technology can be improved by innovations in power transistor design. As previously identified earlier transistors such as JFETs and MOSFETs suffer from on-state losses and reduction in efficiency at higher temperature. These losses are influenced by the material selection for the transistor gate. By facilitating the same voltage but with a thinner gate due to SiC semiconducting properties, on-state internal resistances can be reduced [27]. In conjunction, increased thermal conductance allows SiC MOSFETs to operate with elevated efficiency at higher temperature. Though better than its predecessor, SiC MOSFETs face barriers for mainstream adoptions due to defects seen through manufacturing.

5.5 E-machine optimization

In comparison to inverter and battery design, the fundamental technology of electromagnets are relatively mature for e-machines. Any improvements in e-machine efficiency are most likely to occur with control techniques. One such example is the research and application of switch-reluctance motors. Torque reluctance generated by electromagnets housed in the stator, allow for more simplistic design of the rotor. By doing so, overall rotational inertia can be optimized due to eradication of the electromagnets and commutation present on the traditional designs. Any development in mature products would require considerations of economic feasibility to substitute incumbent products. For motors, this would be the cost of production and materials. Without satisfying this criteria, adoption of new e-machine designs are unlikely.

An overall theme for vehicle energy optimization follows the stream of light-weighting components to reduce overall vehicle inertia. The reduction in either component weight or the number of components can have significant impact on vehicle performance. With a reduced overall mass, less torque is required to achieve the desired acceleration by the driver. Consequently, reduced energy consumption is achieved. With state-of-the-art design software, weight reduction through finite element methods/finite volume methods (FEM/FVM) can remove unnecessary materials from a product. By computing the structural loads going through a block of material, designers can remove excess material from their designs improving the overall weight of the component. This technique could be applied universally through a majority of most structural components of automotive vehicles, alongside developments in manufacturing techniques.

6. Conclusions

In conclusion, vehicle efficiency will be an important factor across all types of vehicles. With limited non-renewable fuel sources, pressure to reduce emissions and importance to bridge the gap between combustion and electric range concerns, energy optimization has become the prime focus to the transport industry. Analyzing the fundamentals of vehicle motion, various characteristic of vehicles can be targeted to improve vehicle efficiency. Reduction of vehicle inertia by weight reduction has been a popular stream that engineers and designers have pursued due to its significance with Newton's equation of motion.

Besides optimization of inertia, overall output of vehicle driving force can be as significant to improve vehicle efficiency. Vehicle drag, a natural commonality across all vehicles has more of a varied significance to a vehicle's efficiency. These loss modes tax vehicle power dependent on driving factors such as vehicle velocity and component temperatures. Optimization of these drag modes are subject to component characteristics.

In contrast, propulsion forces used to propel the vehicle are subjected more down to its efficiency of converting stored energy to useful kinetic energy for motion. As seen between combustion and electric powertrains, the latter method provides power at a greater efficiency. Propulsion methods that rely on combustion to convert stored energy into power waste large portions of energy. This results in 30–40% of actual useful power being used to drive the vehicle.

In comparison, electric powertrains provide tractive power at a high efficiency but rely on vehicle efficiency to maximize range off its limited energy storage. With no significant power conversion between energy mediums, electric powertrains can boast efficiency values of 80–90%. But with larger duration's needed to charge batteries, maximizing efficiency to gain range is currently sought after to ensure electric vehicles are as practical as combustion vehicles.

Thanks

Thank you to the staff and researchers from the Energy systems and Storage team at Warwick Manufacturing Group for the guidance and help provided as I work towards my doctorate.

Abbreviations

MPG	miles per gallon
LHS	left hand side
MTTT	mobile test tire trailer
4WD	four wheel drive
CAD/CAE	computer-aided design/computer-aided engineering
CFD	computational fluid dynamics
WLTP	world light-vehicle test procedure
EV	electric vehicle
OOP	optimum operating point
AC	alternating current
DC	direct current
IGBT	insulated-gate bipolar transistors
MOSFET	metal-oxide-semiconductor field-effect transistor
SOC	state-of-charge
SOH	state-of-health
DoD	depth-of-discharge
SEI	solid electrolyte interphase

Author details

Araan Mohanadass
Warwick Manufacturing Group, University of Warwick, Coventry, United Kingdom

*Address all correspondence to: a.mohanadass@warwick.ac.uk

References

[1] Greene DL, Hopson JL, Li J. Have we run out of oil yet? Oil peaking analysis from an optimist's perspective. Energy Policy. 2006;**34**(5):515-531

[2] How Long Before We Run Out of Fossil Fuels? [Internet]. Our World in Data. Available from: https://ourworldindata.org/how-long-before-we-run-out-of-fossil-fuels [Accessed: 22 May 2016]

[3] Conti J, Holtberg P, Diefenderfer J, LaRose A, Turnure JT, Westfall L. International energy outlook 2016 with projections to 2040. USDOE Energy Information Administration (EIA). Washington, DC, United States: Office of Energy Analysis; 1 May 2016; Section 8:127

[4] Clark SK, editor. Mechanics of Pneumatic Tires. US Government Printing Office; 1981

[5] Lin YJ, Hwang SJ. Temperature prediction of rolling tires by computer simulation. Mathematics and Computers in Simulation. 2004;**67**(3): 235-249

[6] Hall DE, Moreland JC. Fundamentals of rolling resistance. Rubber Chemistry and Technology. 2001;**74**(3):525-539

[7] Sandberg U, Bergiers A, Ejsmont JA, Goubert L, Karlsson R, Zller M. Road surface influence on tyre/road rolling resistance. Swedish Road and Transport Research Institute (VTI), Linkping, Sweden. 2011;**23**:2018

[8] Kluger MA, Long DM. An overview of current automatic, manual and continuously variable transmission efficiencies and their projected future improvements. In: SAE Technical Paper; 1999

[9] Lucente G, Montanari M, Rossi C. Modelling of an automated manual transmission system. Mechatronics. 2007;**17**(2–3):73-91

[10] Roberts A, Brooks R, Shipway P. Internal combustion engine cold-start efficiency: A review of the problem, causes and potential solutions. Energy Conversion and Management. 2014;**82**: 327-350

[11] Howey DA, Martinez-Botas RF, Cussons B, Lytton L. Comparative measurements of the energy consumption of 51 electric, hybrid and internal combustion engine vehicles. Transportation Research Part D: Transport and Environment. 2011;**16**(6): 459-464

[12] Srinivasan R, Baisden AC, Carkhuff BG, Butler MH. The five modes of heat generation in a Li-ion cell under discharge. Journal of Power Sources. 2014;**262**:93-103

[13] Vetter J, Novk P, Wagner MR, Veit C, Mller KC, Besenhard JO, et al. Ageing mechanisms in lithium-ion batteries. Journal of Power Sources. 2005;**147**(1–2):269-281

[14] Spotnitz R. Simulation of capacity fade in lithium-ion batteries. Journal of Power Sources. 2003;**113**(1):72-80

[15] Berecibar M, Gandiaga I, Villarreal I, Omar N, Van Mierlo J, Van den Bossche P. Critical review of state of health estimation methods of Li-ion batteries for real applications. Renewable and Sustainable Energy Reviews. 2016;**56**:572-587

[16] Zhang C, Allafi W, Dinh Q, Ascencio P, Marco J. Online estimation of battery equivalent circuit model parameters and state of charge using decoupled least squares technique. Energy. 2018;**142**:678-688

[17] Schifer R, Lipo TA. Core loss in buried magnet permanent magnet synchronous motors. IEEE Transactions on Energy Conversion. 1989;**4**(2):279-284

[18] Rajapakse AD, Gole AM, Wilson PL. Electromagnetic transients simulation models for accurate representation of switching losses and thermal performance in power electronic systems. IEEE Transactions on Power Delivery. 2005;**20**(1):319-327

[19] Casanellas F. Losses in PWM inverters using IGBTs. IEE Proceedings-Electric Power Applications. 1994; **141**(5):235-239

[20] Jemni MA, Kantchev G, Abid MS. Influence of intake manifold design on in-cylinder flow and engine performances in a bus diesel engine converted to LPG gas fuelled, using CFD analyses and experimental investigations. Energy. 2011;**36**(5): 2701-2715

[21] Abd-Alla GH. Using exhaust gas recirculation in internal combustion engines: A review. Energy Conversion and Management. 2002;**43**(8): 1027-1042

[22] Zheng M, Reader GT, Hawley JG. Diesel engine exhaust gas recirculation: A review on advanced and novel concepts. Energy Conversion and Management. 20041;**45**(6):883-900

[23] Lecointe B, Monnier G. Downsizing a gasoline engine using turbocharging with direct injection. In: SAE Technical Paper; 2003

[24] Falkowski A, McElwee M, Bonne M. Design and development of the DaimlerChrysler 5.7 L HEMI engine multi-displacement cylinder deactivation system. In: SAE Technical Paper; 2004

[25] Fukui T, Nakagami T, Endo H, Katsumoto T, Danno Y. Mitsubishi Orion-MDA new variable displacement engine. SAE Transactions. 1983;**1**: 362-370

[26] Ohta S, Kobayashi T, Seki J, Asaoka T. Electrochemical performance of an all-solid-state lithium ion battery with garnet-type oxide electrolyte. Journal of Power Sources. 2012;**202**: 332-335

[27] Pittini R, Zhang Z, Andersen MA. Switching performance evaluation of commercial SiC power devices (SiC JFET and SiC MOSFET) in relation to the gate driver complexity. In: 2013 IEEE ECCE Asia Downunder. IEEE; 2013. pp. 233-239

Chapter 2

Reliability-Oriented Design of Vehicle Electric Propulsion System Based on the Multilevel Hierarchical Reliability Model

Igor Bolvashenkov, Jörg Kammermann, Ilia Frenkel and Hans-Georg Herzog

Abstract

This chapter describes a methodology of evaluation of the various sustainability indicators, such as reliability, availability, fault tolerance, and reliability-associated cost of the electric propulsion systems, based on a multilevel hierarchical reliability model (MLHRM) of the life cycles of electric vehicles. Considering that the vehicle propulsion systems are safety-critical systems, to each of their components, the strict requirements on reliability indices are imposed. The practical application of the proposed technique for reliability-oriented development of the icebreaking ship's electric propulsion system and the results of computation are presented. The opportunities of improvement of reliability and fault tolerance are investigated. The results of the study, allowing creating highly reliable electric vehicles and choosing the most appropriate traction electric drive design, are discussed.

Keywords: electric vehicle, reliability-oriented design, fault tolerance, electric propulsion system, multilevel hierarchical reliability model, Markov model

1. Introduction

The rapid modern development of new technical systems in various areas of the industry is directly related to a significant increase in their complexity. In addition, the levels of integration of subsystems, units, and components and, accordingly, their mutual effect largely increase as well. This, in turn, has a very strong impact on the reliability, fault tolerance, and maintainability of the designed technical systems. Reliability concepts can be applied to virtually any engineered system. In its broadest sense, reliability is a measure of performance.

All of the above fully applies to the traction drive of electric vehicles, the creation of which is a major challenge in the modern way to the electrification of the different types of vehicles: ships, planes, trains, helicopters, busses, and cars. For transport facilities that are safety-critical systems, the issues of assessing and optimizing reliability indicators are of particular importance.

As can be seen in **Figure 1**, the magnitude of the level of technical excellence of an electric traction drive is determined by three comprehensive criteria: sustainable functioning, efficient functioning, and environmental level. It follows in **Figure 1**

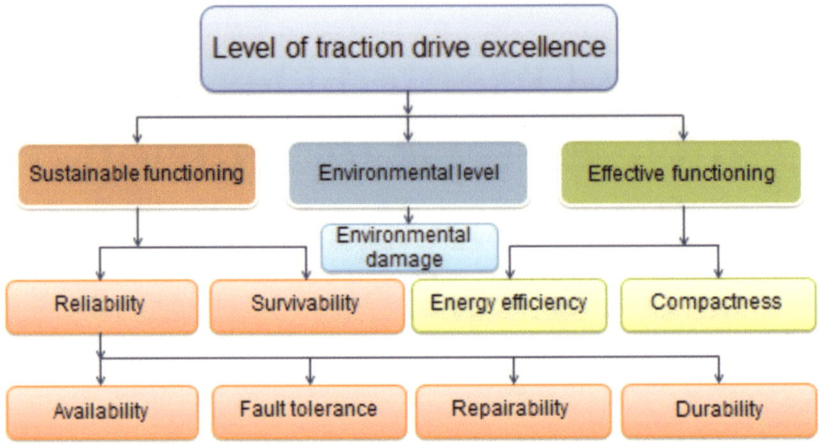

Figure 1.
Structure of the traction drive level of excellence.

that the maximum number of factors affects the amount of sustainable functioning criterion of the traction drive. Accordingly, the above criterion has the maximum potential to increase the value of the level of excellence of the traction electric drive and an electric vehicle as a whole. In addition, the most stringent requirements are imposed on reliability, fault tolerance, and survivability of electric vehicles, which are safety-critical systems.

In this way, reliability-oriented design of the vehicle electric propulsion system and, accordingly, all its subsystems, units, and components is a very urgent and complex task while considering their interactions. In recent years, a multilevel approach in the development, design, and optimization of various technical systems and their particular parameters has become quite widespread. In addition, when using a multilevel approach in most cases, the various levels are interconnected hierarchically. Depending on the complexity of the system being developed, the multilevel hierarchical reliability model (MLHRM) may consist of a different number of levels. In the simplest case, it can consist of three levels.

Attempts to develop the methods for solving such a problem were undertaken by various research groups. The first group of scientists, whose works are presented in [1–4], uses the method of hierarchical decomposition of the technical system, better known as analytic hierarchy process (AHP). It was developed by Thomas L. Saaty in the 1970s and represents a structured technique to organize and analyze complex decisions, described in detail in [1]. This approach has significant advantages when important components of the decision are difficult to quantify or to compare or when communication between team members is made difficult by their different specializations, terminology, or perspectives. Due to the relatively simple mathematical formula, as well as the easy data collection, AHP has been widely applied by many researchers. The integral shortcoming of the AHP is the fact that the criteria are assumed to be completely independent, even though in real-world problems, the criteria are often dependent. In [2] the AHP approach was applied in the four-level hierarchical tree to identify the main attributes and criteria that affect the level of accuracy of the models used in probabilistic risk assessment. The main disadvantage of AHP approach is the inability to consider the uncertainties of the process. In order to overcome this limitation, the application of different hybrid combinations of fuzzy theory and AHP, the so-called fuzzy AHP, and analytic network process (ANP) method has been used in [3] for inter-criteria dependency definition and in [4] for the vehicle safety analysis. It should be noted that in real life, most of the decision problems are represented by a network and not only structured as a hierarchy.

Various hierarchical stochastic models have proven to be a powerful tool for analyzing the reliability of complex technical systems for different applications. The authors in [5] described a method, called the hierarchical Markov modeling (HMM), which allows to perform the predictive reliability assessment of distribution electrical system. This method can be used not only to assess the reliability of existing distribution systems but also to estimate the reliability impact of several design improvement features. HMM creates a primary model based on the system topology, secondary models based on integrated protection systems, and tertiary models based upon individual protection devices. Once the tertiary models have been solved, the secondary models can be solved. In turn, solving the secondary models allows the primary model to be solved and all of the customer interruption information to be computed. An interesting approach to solving the complex problem of performance, availability, and power consumption analysis of infrastructure as a service (IaaS) clouds, based hierarchical stochastic reward nets (SRN), is presented in [6]. In order to use the resources of an IaaS cloud efficiently, several important factors such as performance, availability, and power consumption need to be considered and evaluated carefully. The estimation of these indicators is significant for cost–benefit prediction and quantification of different strategies, which can be applied to cloud management.

Possible techniques and ways to solve the problem of a multistage reliability-based design optimization (MSRBDO) are based on Monte Carlo method and its application to aircraft conceptual design, which is described in detail in [7] and with subsequent corrections and development in [8]. In recent years, a multilevel (tiered) systematic approach has become increasingly widespread for analyzing and optimizing the various characteristics of technical systems, the theoretical foundations of which are described in detail in [9–12]. In the work of [9], the four-level (system, subsystem, assembly, and device-component) representation of variable-speed drive systems is proposed for the analysis of reliability, availability, and maintainability. The calculations were performed analytically and step by step. Bolvashenkov et al. [10] describes the rules and properties of multilevel hierarchical representation of the vehicles' propulsion system life cycles and the optimal types of stochastic methods and models for use at each individual level. A new look at solving the problem of assessing various system resilience, based on the three-level (tiered) approach, is proposed in [11]. Ref. [12] presents a systematic four-level approach to develop the reliability design of the mechanical system—the refrigerator, which is similar to the target of this chapter, but it does not present any analytical optimization.

A significant amount of research works is related to the assessment of the reliability of particular units or component at one of the local levels of the multilevel model and the development of appropriate methods and models [13–16]. In Refs. [13, 14], several options for assessing reliability at the component level are presented. In the first case [13], it is proposed to do this using failure mode and effect analysis (FMEA) with weighted risk priority number (RPN), and in the second case [14], it is proposed to do this based on a multistate Markov model, which allows to consider random environmental conditions. The hierarchical model for lithium-ion battery degradation prediction, discussed in [15], represents reliability assessment technique at the unit level of a multilevel model. The three-level (system, subsystem, and component) aircraft engine model's hierarchical architecture is described in [16]. This paper concludes that in a large system, such as an aircraft engine, failure prognostics can be performed at various levels, i.e., component level, subsystem level, and system level. A similar approach for the estimation of the remaining useful life (RUL) for the multiple-component systems—when using the prognostics and health monitoring (PHM) technologies in modern aircraft—is proposed in papers [17, 18]. This methodology combines particular component RUL estimations into a single system level RUL estimation. This characteristic becomes more relevant when the number of components within the system increases.

2. Methodology of a multilevel hierarchical reliability model

In order to solve the problem of implementing the reliability-oriented design for electric propulsion system, the authors, based on previous own research and research of other scientists, developed the methodology for creating and using the MLHRM of electric vehicles' functioning. The main features, techniques, and potentials of the model are presented below.

The proposed method of reliability-oriented design of the vehicle electric propulsion system based on the MLHRM allows to solve a complete set of tasks related to the full range of indicators of comprehensive reliability for the safety-critical electric traction systems, such as failure-free operation probability, fault tolerance, availability, maintainability, durability, reliability associated cost, etc.

The main advantages of the proposed methodology derive from the use of system approach principles for the development of the methodology and the bidirectional principle of the MLHRM functioning. In accordance with the principles of the system approach, the model allows to take into account the horizontal and vertical interaction of components of different levels of the MLHRM, considering the impact of the real operating conditions.

The bidirectional structure of the model functioning allows to solve the problems of reliability and fault tolerance optimization of electric vehicles, both at the stage of designing and in the stage of operation.

2.1 Structure of MLHRM

Figure 2 shows the general view of the MLHRM structure. The number of levels of the model can vary depending on the complexity of the technical system and the tasks to be solved. The model presented in **Figure 2** has six levels, which correspond

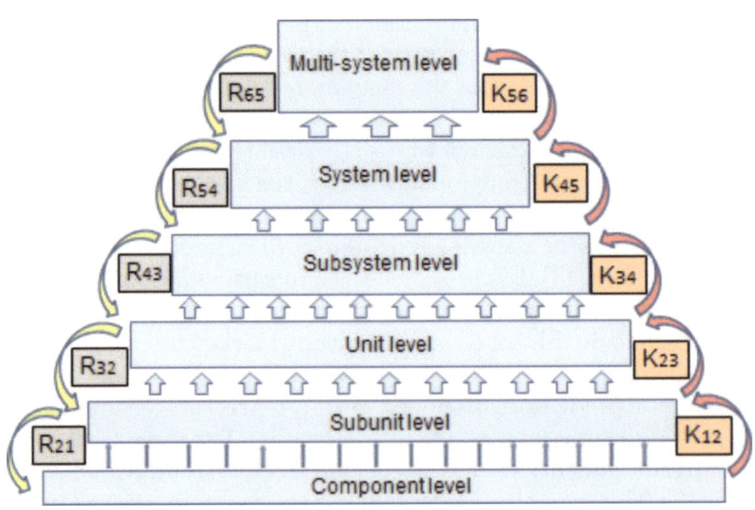

Figure 2.
General structure of the MLHRM.

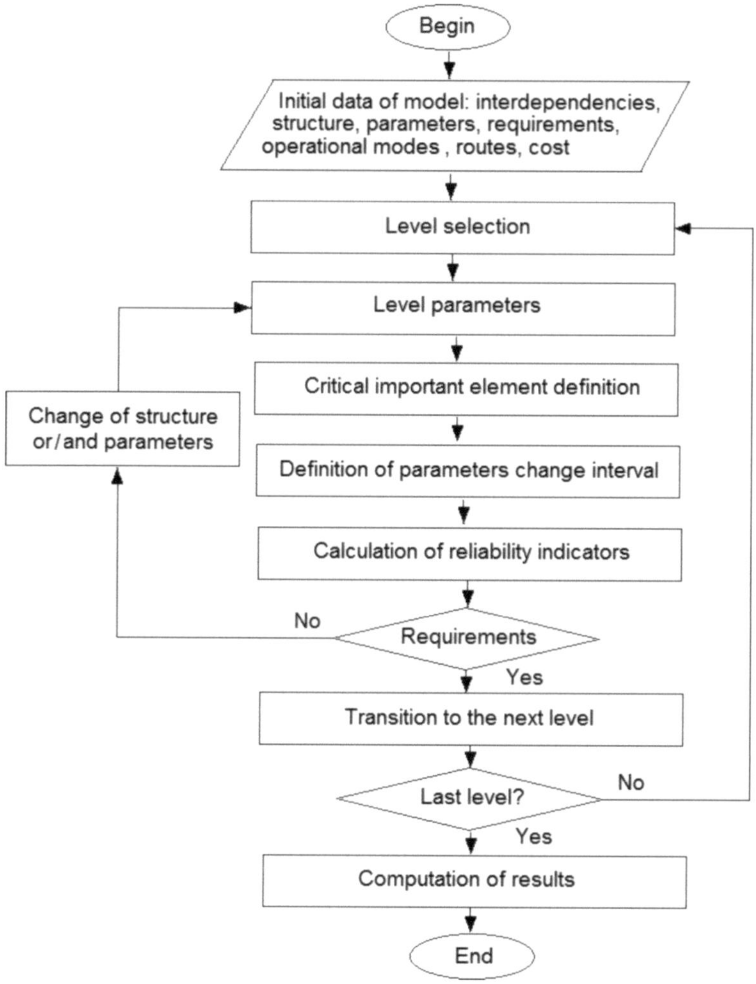

Figure 3.
Algorithm for rapid analysis of the reliability characteristics of a technical system.

to the task of analyzing and optimizing the reliability characteristics of electric vehicles, taking into account their interaction in random environment.

The coefficients K_{12}–K_{56} determine the magnitude of the influence of the reliability of the lower level of the model on the neighboring upper level. The coefficients R_{21}–R_{65} determine the ratio of the required values of the performance of the upper level of the model relative to the neighboring lower level.The coefficients K_{12}–K_{56} from **Figure 3** can be defined by Eq. (1):

$$K_{n(n+1)} = \Sigma(Cr_{ni} \cdot P_{ni}), 1 \le n \le 6, 1 \le i \le m_n, \quad (1)$$

where Cr_{ni} is the criticality value of the ith element of the nth level, P_{ni} is the failure probability of ith element of the nth level, and m_n is the number of elements of the nth level.

The coefficients R_{21}–R_{65} can be computed by Eq. (2):

$$R_{(n+1)n} = Y_{(n+1)}/X_n, 1 \le n \le 6, \quad (2)$$

where $Y_{(n+1)}$ is the upper neighboring level performance, X_n is the lower neighboring level performance, and n is the number of level.

The coefficients R_{21}–R_{65} are used to calculate the required indicators of various levels of the MLHRM within the design of electric vehicles with the specified reliability and fault tolerance parameters.

The coefficients K_{12}–K_{56} are used to improve the reliability indicators of various levels of the model during the operational time of the electric vehicles.

As noted above, the MLHRM shown in **Figure 2** includes six levels, namely, component level (CL), subunit level (SUL), unit level (UL), subsystem level (SSL), system level (SL), and multi-system level (MSL).At the CL, based on statistical reliability data, analytical calculations, or using Markov models for binary-state components, reliability characteristics of the element of the next level (SUL) are determined. In operational mode, component failures can lead to the degradation of the whole system performance. Respectively, the performance rate of any component can range from fully functioning up to complete failure. The failures that lead to a decrease in the element performance are called partial failures. After partial failure, the elements continue to operate at reduced performance rates, and after complete failure, the elements are totally unable to perform their missions.

At the SUL the initial parameters for the analysis of reliability indicators of the red level are determined. As subunits, the independent functional parts of the next level (UL) can be considered. In turn, at the UL, an analysis and evaluation of independent functional units, which are integral parts of the next level, SSL, are carried out.

The reliability indicators calculated at the UL are the input data for the models used within the next level—the SSL. In the case of electric vehicle simulation, the SSL corresponds to the level where the assessment of the reliability characteristics of the entire electric traction drive takes place. The basic model of the vehicle electric propulsion system at this level can be represented as stochastic model of multistate system with the change of discrete operating load modes. Each operational load mode complies with specific power characteristics, which have to be implemented with highest probability for safety operation of the vehicle. Thus, on the one hand, there are requirements for safe vehicle operation, which form a model of demand. On the other hand, there is the guaranteed generated electric power, which values form the model of performance. The combined performance-demand model allows to determine the characteristics of reliability, based on which it is possible to estimate the degree of fault tolerance of the vehicle's electric propulsion system and to optimize its values according to the project requirements.

At the SL, complex reliability indicators of electric vehicle are investigated. The input data for modeling at this level of the MLHRM are the output reliability characteristics, which are obtained at the SSL. In turn, the output characteristics of SL are the input data for models of the top-level MSL. At the MSL, the reliability-associated economical characteristics of the joint operation of a multiple number of electric vehicles under real operating conditions are estimated taking into account their interaction and random environment. The problems solved at this level were not the purpose of the present study and, therefore, are not considered in this chapter.Based on the presented MLHRM, an algorithm was developed for the accelerated estimation of the compliance of the propulsion system reliability indicators with the project requirements, which is shown in **Figure 3**.

In accordance with the above algorithm, the main task of a simplified rapid assessment of reliability indicators is to determine the critical important components of each level of MLHRM and the degree of its influence on the reliability characteristics of the neighboring upper level.

In this case, the critical important parts of each level can be determined based on risk priority number (RPN), failure mode and effects and criticality analysis (FMECA) or based on experimental data, as shown in **Figure 4**, which was previously presented in [19–21] for the main subunits of the traction electric motor: stator windings, power electronics, and bearings.

Depending on the task to be solved and the level of the model, the probability of failure-free operation, availability, degree of fault tolerance, etc. can be considered as indicators of reliability of the components.

In order to meet the requirements of the project on reliability and fault tolerance of electric vehicles, it may be necessary to change the reliability parameters of the components and/or the structure of the electric propulsion system.

The intervals of possible changes in the reliability parameters (failure rate, repair rate) of the propulsive system elements are determined preliminarily based on statistical data on the reliability of each element, given, for example, in the reference literature.

From the results shown in **Figure 4**, it follows that the most sensitive parts to thermal effects in various operating conditions and in terms of reliability are the stator windings of the traction electric motor. In this case, for further investigations, the stator windings are accepted as a critical important subunit for the unit—the traction electric motor. Similarly, the critical important parts for the remaining levels of MLHRM can be defined.

2.2 Goals, methods, and models

At each level of the MLHRM, specific models are used to solve specific tasks in order to achieve the corresponding goals at each level. **Figure 5** graphically presents the problems associated with the reliability characteristics of electrical propulsion systems that can be solved by means of the MLHRM. In addition, **Figure 5** presents the methods and models recommended in order to assess the reliability indicators of different MLHRM levels.

Below, a detailed description of the tasks and methods for their solution, applied to each level of MLHRM, is given.

2.2.1 Component level

The main tasks that are solved at the CL are the collection, analysis, and structuring of statistical data on the reliability of all components that affect the reliability of the neighboring top level of the MLHRM. It also identifies the critical

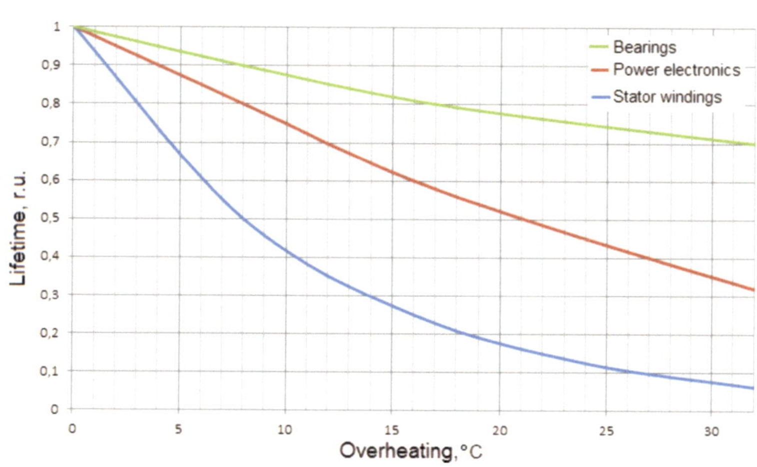

Figure 4.
Critical importance analysis of the subunits [19].

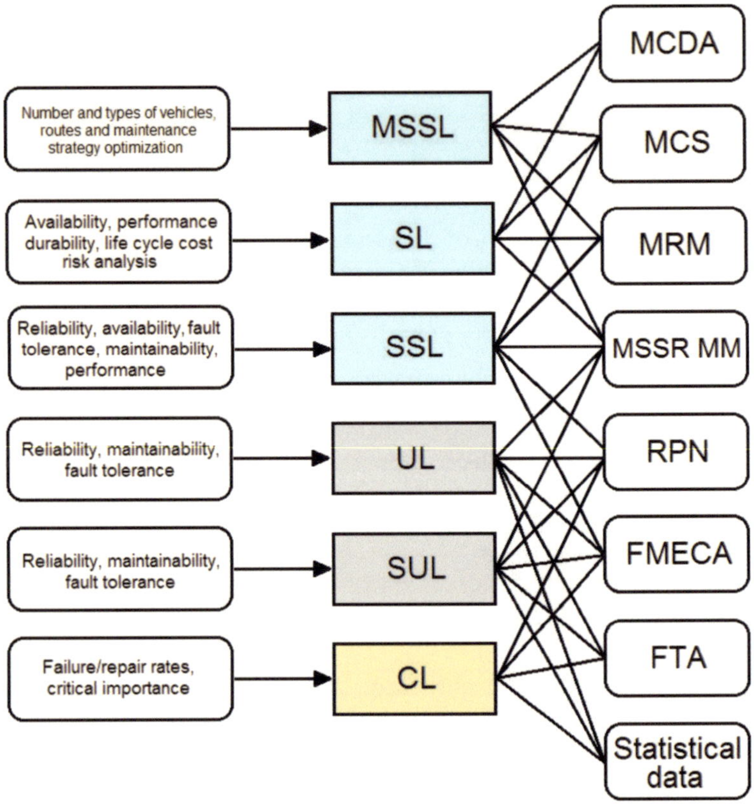

Figure 5.
Tasks and methods of their solutions for different MLHRM levels.

important components and their degree of influence on the reliability features of the next level—the SUL. The possible methods for achieving these goals are fault tree analysis (FTA), failure mode and effects analysis (FMEA), FMECA, and RPN. Several examples of the reliability characteristic analysis of electric propulsion systems at CL of the MLHRM are described in [21–23].

2.2.2 Subunit level

As subunits, this chapter examines individual, relatively independent parts of units having a specific functional orientation. At the subunit level, based on the data obtained in the previous component level, it is advisable to determine the characteristics of reliability, maintainability, and fault tolerance of the subunit groups, forming the corresponding elements of the next level—the UL. The recommended methods for analyzing and evaluating the above reliability characteristics are FTA, FMEA, FMECA, and RPN using experimental failure and repair statistics. If there are blocks that are not binary, but multistate elements (elements with degraded states), the multistate system reliability Markov models (MSSR MM), described in detail in [20, 23, 24], can be applied for the computation.

2.2.3 Unit level

At the UL, the tasks of computation and optimization of reliability, maintainability, and fault tolerance of autonomous functional parts (units), within the propulsion system of electric vehicles, are solved. Taking into account that the units are elements with several degraded states, that is, multistate systems, it is advisable to use MSSR MM for their research. In addition, by means of MSSR MM, one can take into account the actual load modes of the units, regarding overloads capacity and the aging processes. The transition probabilities for MSSR MM can be calculated by means of the degree of fault tolerance DOFT [24] using statistical operational data or can be determined at the design stage based on the requirements to the safety and sustainable vehicle operations. In order to determine the critical important elements of the UL for further optimization, RPN, FMECA, FTA, and experimental test methods can be used.

2.2.4 Subsystem level

At the SSL the problems of determining and optimizing the reliability characteristics of operational availability, maintainability, fault tolerance, redundancy (functional and structural), and performance of entire electric propulsion system should be solved. In order to build the corresponding combined stochastic model of the electric vehicle propulsion system including electric energy source, the concept of balanced relationship between demand (required power) and performance (available power) has been applied. Hence, the model of the electric propulsion system operation can be represented as a MSSR MM with the change of discrete operating modes: start (takeoff), acceleration (climb), constant speed (cruise), deceleration (reduction of altitude), and stop (landing). Along with MSSR MM, Markov reward models (MRM) and Monte Carlo simulation (MCS) can be widely apply.

2.2.5 System level

At this level, the most preferred are the various stochastic models of the electric vehicle's lifecycle, which allow to assess the reliability indices of repairable systems by optimizing maintenance strategies according to the intensity of the scheduled and unscheduled repairs, and the use of functional systems of monitoring, forecasting reliability, and diagnostics. These may be MSSR MM, MRM, MCS, and multi-criteria decision analysis (MCDA). A definition of current and forecasted values of reliability indices are carried out, considering the external and internal operation conditions of the vehicle, as well as taking into account the availability of structural or functional redundancy. Thus, the study and optimization task of the so-called reliability associated costs (RAC) estimation, based on MRM, is most interesting and promised [20].

In order to build such a model, the process of the vehicle operations can be represented by a chain of the lifecycles: operational, nonoperational, working, standing, etc. The data on the duration of each cycle are obtained based on the analysis of statistical operational data of a particular type of vehicle on certain routes and areas.

3. Application case

As an application example of the proposed MLHRM methodology for assessing and optimizing the reliability characteristics of electric traction drives, the propulsion system of icebreaking cargo ship is considered. Functionally, the MLHRM is presented in **Figure 6**. The new Arctic liquefied natural gas (LNG) tanker "Christophe de Margerie," built in 2017 by Daewoo Shipbuilding & Marine Engineering in South Korea, was selected as the research object to investigate the reliability features of the overall electric propulsion system. The characteristics of the LNG tanker "Christophe de Margerie," as well as its propulsion system are described in detail in [25].

Reliability indicators of lower levels have been calculated based on statistical data (failure rates, repair rates, etc.), and well-known analytical methods are not included to this chapter, however, are fully presented in [25]. This chapter concentrates on the upper levels, which are more complicated and interesting considering the overall electric vehicle reliability.

As a MSL of the MLHRM in this case, the joint operation of several ships in a caravan with icebreakers, the joint operation of the whole fleet to deliver the similar type of cargo in corresponding directions, the operation of the shipping company, etc. can be considered.

In **Figure 6**, the following notation is used: EES, electric energy source; EC, electric converter; EM, electric motor; CU, control unit; and λ_j, λ_k, λ_n, and λ_m, failures rates of various components.

The main goal of the ship's propulsion system is to ensure the safe and efficient transportation of cargo and/or passengers. Based on the stated main goal, the functions that should be performed at each level of the MLHRM are analyzed. Below is a detailed description of each model level applied to the ship's electrical propulsion system. For a more complete understanding of the essence of the multilevel structure of the MLHRM, **Figure 7** shows the most simplified diagram of the fully integrated power system of the icebreaker LNG tanker.

The entire ship's power system can be conventionally represented as three subsystems: the electric energy source system (EES), the ship's electric propulsion system (EPS), and the subsystem of the ship's consumers of electric energy (EEC). The first subsystem includes six diesel generators with a total power of 62 MW, which supply electric energy to a two-section main switchboard.

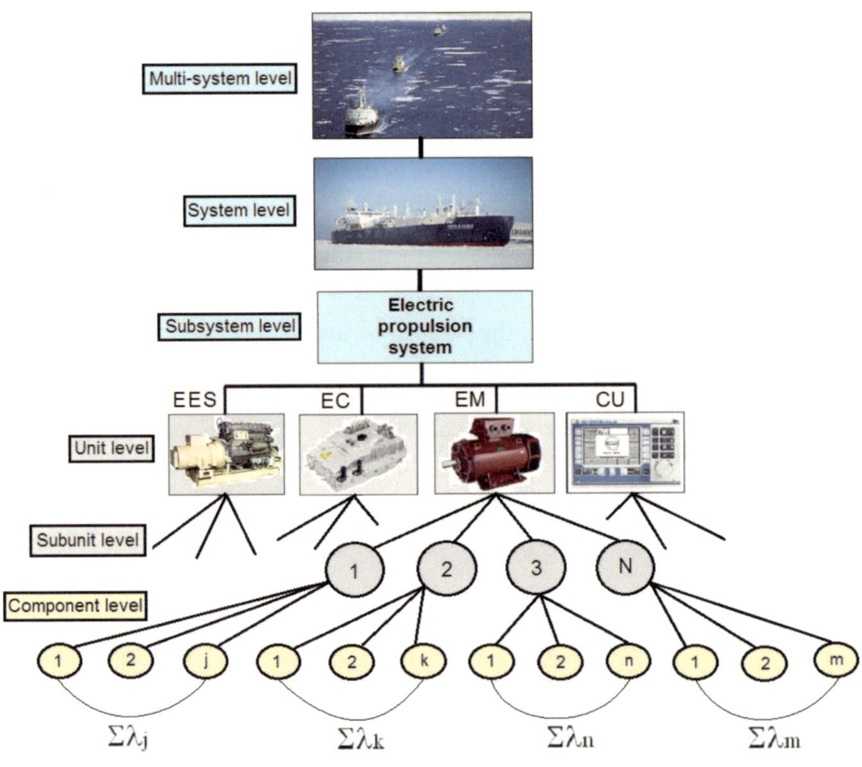

Figure 6.
MLHRM structure of icebreaking cargo ship with electric propulsion.

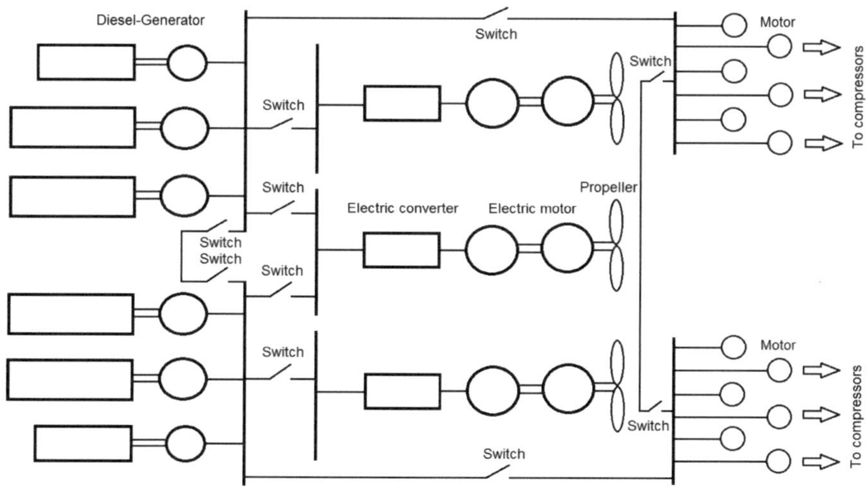

Figure 7.
Structure of the whole power system [25].

The electric propulsion subsystem consists of three electric traction drives, including electric converters and three two-section electric traction motors, located in steering gondolas of the Azipod system. The ship's consumer subsystem provides general ship needs, as well as the critical important consumer, namely, the gas liquefaction and storage system (LSS), consisting of 12 powerful motor compressors.

When transporting LNG, specifically stringent requirements are imposed on the whole power system of the tanker in terms of safe and sustainable operation. On the one hand, in the heavy ice conditions of the Arctic, it is necessary to ensure the maximum possible power on all three propellers of the vessel, and on the other hand, in the same time, it is necessary to ensure uninterrupted functioning of the LSS for the safety and keeping of the cargo. This feature should be unconditionally observed during the simulation on SL and MSL. It should be noted that this requirement extends over 50% of the operating time of LNG tanker.

3.1 Component level and subunit level

At the component level, based on available failure statistics [21–23] and the above methods of analytical reliability calculation (FTA, FMEA, RPN, etc.), the total failure rates of all components, of which the subunits are composed, can be analyzed and estimated. For EM, as the part of UL, the subunits are a stator with windings, a rotor with magnets, a bearing, and others, as shown in **Figure 8**.

Considering the above data in **Figure 8**, generally the reliability of electric motor λ_{EM} can be determined by the formula:

$$\lambda_{EM}(t) = \Sigma\lambda_{Si}(t) + \Sigma\lambda_{Rj}(t) + \Sigma\lambda_{Bk}(t), \quad (3)$$

where λ_{Si}, λ_{Rj}, and λ_{Bk} are the failure rates of parts of all parts of the electrical machine, respectively, of stator, rotor, and bearing.

For EC, as the part of UL, the subunits are the semiconductors, printed circuit boards (PCB), capacitors, and others, as shown in **Figure 9**.

Based on the above data in **Figure 9**, generally the failure rate of an electric power converter λ_{EI} can be estimated considering the reliability values of its components by the equation:

$$\lambda_{EC}(t) = \Sigma\lambda_{Ti}(t) + \Sigma\lambda_{Dj}(t) + \Sigma\lambda_{Ck}(t) + \Sigma\lambda_{Bn}(t), \quad (4)$$

where λ_{Ti}, λ_{Dj}, λ_{Ck}, and λ_{Bn} are the failure rates of all components of electric inverter, respectively of transistor, diode, capacitor, and printed circuit board.

Similar calculations are performed for all other subunits of the SUL, which are taken into consideration. Based on the results of the calculation, the sensitivity of changing the values of the reliability indicators at the subunit relatively to the change of the components' failure rates is determined. The obtained results are used further in the models at UL and SSL.

Increased reliability features on the CL can be performed using components and materials with higher reliability values and based on various methods of critical components redundancy. In order to achieve the required performance characteristics of the SUL, as shown in [21], it is necessary to optimize the type of stator windings, permanent magnets, bearings, semiconductors, etc. In addition, redundancy of critical important parts of subunits can be used.

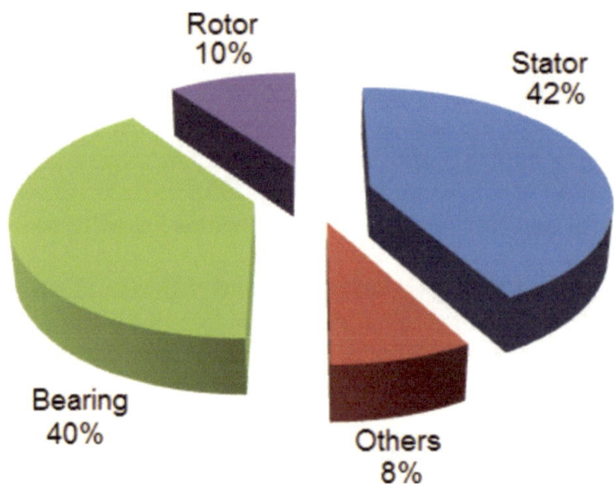

Figure 8.
Failures statistics of traction electric motor [21].

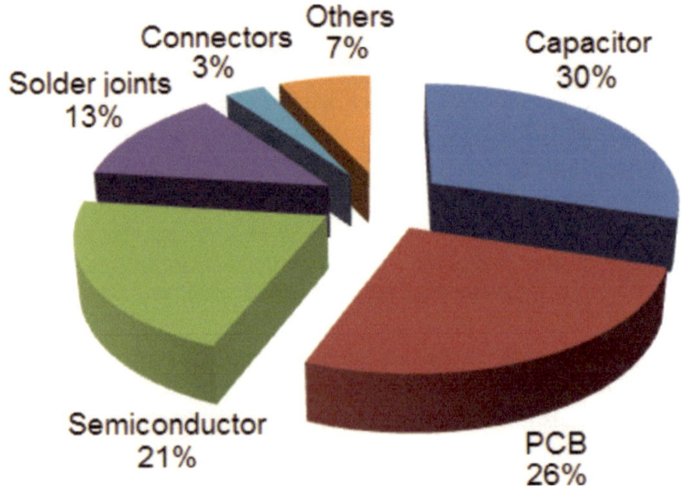

Figure 9.
Failures statistics of electric converter [21].

3.2 Unit level

At this level of the MLHRM, the tasks of providing reliable performance of all functional elements, which form the subsystem of the electrical propulsion system presented in **Figure 6**, are solved. The detailed descriptions of the use of various techniques to improve the reliability and fault tolerance of electric energy sources, traction electric motors, electric converters, and control units at this MLHRM level are given in [19, 20, 23, 26].

The correct choice of the type of electric machine, the methodology of which is presented in [21], has a significant impact on the reliability indicators of an electric propulsion system. Based on the completed studies, it was proposed to use a synchronous motor with permanent magnets as the most promising one in terms of reliability and fault tolerance. One of the most effective methods to improve the reliability and fault tolerance of traction electric motors is the use of a multiphase motor topology with concentrated windings and galvanically uncoupled phases, described in [19, 26]. A significant influence on the characteristics of fault tolerance and overload capacity of the traction electric motor is provided by the parameters and the location of the permanent magnets on the rotor. In the work of [21], it is shown that the most preferable design is the permanent magnet synchronous motor with internal v-shaped arrangement of permanent magnets on the rotor.

The methods to analyze and improve the reliability of the electrical energy source and of the electric converter are discussed in [23, 27]. In order to meet the design requirements for reliability and fault tolerance as shown in [23], as electric energy sources, it is advisable to apply the energy storage, with a matrix topology of battery or fuel cells with more than 20% cells redundancy. Additionally, in order to improve the fault tolerance of the electric power converter in failure cases, it is proposed to use a multilevel cascaded converter topology. The reliability characteristics of all units, taking into account the specific load conditions and aging processes, are advisable to be computed by means of the MSSR MM, as shown in [20, 21, 23, 24].

3.3 Subsystem level

At this level, the entire spectrum of technical tasks, which are related to the most important subsystem of an electric vehicle, is solved. The results of solving these problems will allow at higher levels to determine the financial equivalent of an important indicator of the level of excellence of an electric propulsion system—the sustainable functioning. Such tasks include the analysis and optimization of reliability, operational availability, fault tolerance, maintenance strategies, reliability associated cost, and performance of the propulsion system.

When analyzing the reliability characteristics at the SSL, it is necessary to take into account the operational load modes, the mutual influence between the units, the aging processes, the frequency, and the duration of maintenance and repairs, as well as the influence of structural and functional redundancy of the entire subsystem or its particular parts. The required degree of redundancy of the electric propulsion system of the icebreaker LNG tanker, depending on the requirements for the safety and fault tolerance, can be achieved on the SSL by using multi-power electric energy sources (MPEES) consisting of six diesel generator sets. The questions of features and the analysis of the reliability characteristics of MPEES are described in detail in [27, 28].

High survivability and fault tolerance of the electric propulsion system of LNG tanker are especially important in the extremely difficult ice conditions of the Arctic. In order to ensure the safe and sustainable navigation in the ice conditions, on the SSL, it is necessary to provide the multi-motor electric drives with multiphase electric motors, whose features are discussed in [27, 29].

The most comprehensive investigation of reliability indicators at the SSL is advised to be carried out by means of MSSR MM, MRM, and MCS. Moreover, taking into account the high complexity of Markov models with a high number of states for the entire electric power system, it is proposed to perform the calculations using the new powerful Lz-transform method, described in detail in [20], which drastically simplified the solution of multiple differential equations.

3.4 System level

At the SL, the operation of the ship with electric propulsion subsystem as a whole system is considered. The objective function of the icebreaker LNG tanker is the safely, sustainable, and efficient shipping in the specified Arctic operating conditions. In accordance with this, the main objectives are to increase the carrying capacity of the tanker and to minimize the total operating costs and damages. The reliability characteristics of the icebreaker LNG tanker influence the values of both components of the objective function of the ship. In order to solve these problems, it is advisable to use MCS and MCDA, considering the random environment of the Arctic navigation conditions and the number of uncertainties, along with MSSR MM and MRM.

In this way, at the SL, it is recommendable to determine all reliability indicators of the whole tanker. Based on such reliability indices, the total cost can be calculated, which is needed to maintain sustainably the required level of performance during the operation of the tanker in real ice operating conditions. These are the operational availability, performance, deficiency of performance, maintainability, reliability associated cost, damages from unreliability, life cycle cost, risk probability, etc.

In order to improve the reliability and fault tolerance of the electric propulsion system and the LNG tanker as a whole, at this level, it is possible to use several autonomous electric drives with their own screws, the propulsion system of gondola type with two screws, the optimization of the maintenance and repair strategy of the power system of the tanker during navigation, predictive reliability monitoring, and a control system of the ship electrical propulsion system.

In order to build the model of the LNG tanker life cycle at the SL, the process of the icebreaker LNG tanker operations is represented by a chain of different operating modes. During the operation cycle depending on conditions of navigation, it is possible to distinguish four basic operating modes of an icebreaker LNG tanker. Each of them corresponds to a certain required number and power of the main engines. These operating modes are shown in **Figure 10** and they are:

- Loading and unloading of LNG at the terminal. Each of these two modes usually takes about 24 h. The sustainability of the loading and unloading process is determined by the reliability of onshore and ship gas liquefying and pumping systems.

- Navigation of a ship in ice-free water. The operation in this mode depends on the required velocity and needs of the greater part of the operational time 50–80% of the nominal generated power.

- Autonomous movement in the ice without icebreaker support. The navigation in this mode depends on ice conditions and a wide power range from 50% up to 100% of the nominal power can be used.

- Navigation of a ship in heavy ice supported by icebreakers. In order to realize sustainable joint operation with icebreakers in this mode, electric propulsion system needs 80–100% of the nominal generated power.

Considering the abovementioned features of operational modes of the icebreaker LNG tanker propulsion system, three demand levels were chosen for calculation: 100, 80, and 50% of the main traction electric motors power.

For an accurate assessment of operational availability and performance of the electric propulsion system, it has been proposed to estimate the values separately for each of the above modes, followed by calculating the total impact on the value of the ship's operating speed and, accordingly, the amount of cargo transported per unit of time.

In order to analyze the reliability indicators at the system level of the MLHRM, the icebreaker LNG tanker power system—based on the decomposition principle—is presented in the form of four blocks: the electric energy source system (EES), the ship's electric propulsion system (EPS), the subsystem of the ship's consumers of electric energy (EEC), and LNG liquefaction and storage system (LSS). The simplified structure of the whole LNG tanker power system is shown in **Figure 11**.

As a result of calculating the comprehensive reliability indices of each functional block, indicated in **Figure 11**, based on the Lz-transform method [25, 29] to solve the system of differential equations of MSSR MM, a schedule of operational availability of the power system of LNG tanker for different demands was constructed, which is presented in **Figure 12**.

The graph in **Figure 12** demonstrates the ability of the tanker's power system to ensure sustainable functioning under the conditions of various operational

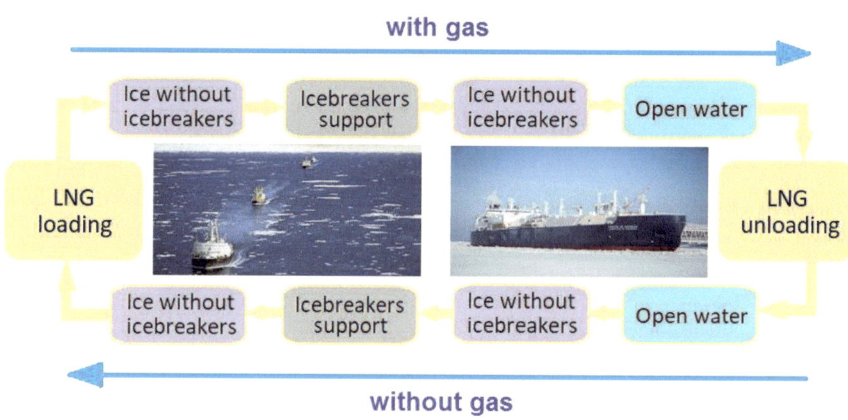

Figure 10.
Operational modes of icebreaker LNG tanker.

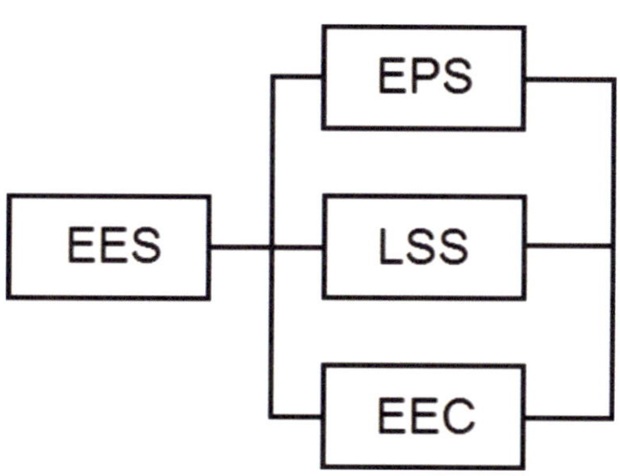

Figure 11.
Structure of the hybrid-electric power system of LNG tanker.

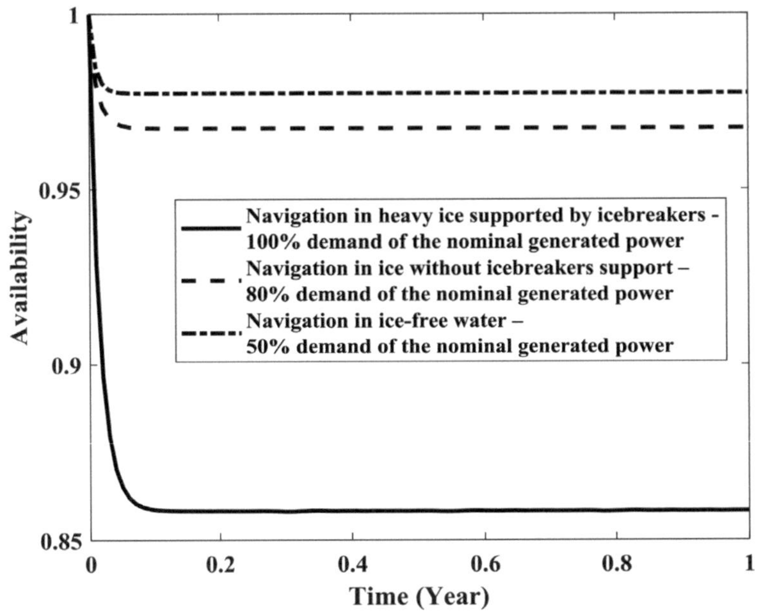

Figure 12.
Operational availability of the power system of LNG tanker for different demands [25].

demands. For this, the process of operating a fully loaded tanker during LNG delivery from the Sabetta terminal on the Russian Yamal Peninsula to the Chinese port of Shanghai was modeled. As can be seen in **Figure 12**, the Arctic LNG tanker has high operational availability for the maximum levels of demand. Its value is equal to 85.82%. This indicates that such multi-drive propulsion system is closely related to the conditions of ice navigation.

4. Conclusions

The chapter proposed MLHRM and methodology of its application will allow to realize the comprehensive analysis an estimation of comprehensive reliability characteristics of the vehicle electric propulsion systems at the design stage. This means to implement the so-called reliability-oriented design of the traction electric drives. The suggested MLHRM of the vehicle's life cycle allows for each level to solve specific technical and technical-economical optimization tasks, such as the optimization of the design of the electric machine, number of phases, number of electric motors, degree of fault tolerance, level of redundancy, maintenance strategy, topologies of electric converters, and electric energy sources.

The MLHRM approach allows to provide a quantitative comparative analysis of methods for improving the comprehensive reliability of the vehicle electric propulsion systems at each MLHRM level. In other words, in order to quantify the impact on the integrated reliability of the electric propulsion system and vehicle as a whole, it is possible to use systems of diagnostics, fault detection, monitoring, fault prediction, varying degrees of redundancy of elements, and various maintenance strategies.

As the application case, the new Arctic LNG tanker "Christophe de Margerie" is used to assess the value of the operational availability of the integrated electric power system during the summer-autumn period along the Northern Sea Route. The results of the research showed that regarding the sustainable operation during

Arctic navigation of the icebreaking LNG tanker, the electric propulsion system has a significant potential to improve operational availability, technical performance, and consequently economic efficiency.

For further studies, it is advisable to estimate the value of the reliability-associated costs, as well as life cycle costs of Arctic LNG tanker for different operational routes by using different maintenance strategies, considering the gradual deterioration of the ship's icebreaking capacity during ice navigation.

Author details

Igor Bolvashenkov[1*], Jörg Kammermann[1], Ilia Frenkel[2] and Hans-Georg Herzog[1]

1 Institute of Energy Conversion Technology, Technical University of Munich (TUM), Munich, Germany

2 Center for Reliability and Risk Management, SCE, Beer Sheva, Israel

*Address all correspondence to: igor.bolvashenkov@tum.de

References

[1] Bani-Mustafa T, Pedroni N, Zio E, Vasseur D, Beaudouin FA. Hierarchical tree-based decision making approach for assessing the trustworthiness of risk assessment models. In: Proceedings of the International Topical Meeting on Probabilistic Safety Assessment and Analysis (PSA'17); 24th–28th September 2017; Pittsburgh, PA. 2017. pp. 314-323

[2] Saaty TL. Decision making with the analytic hierarchy process. International Journal of Services Sciences (IJSSci), Inderscience Publishers. 2008;**1**(1):83-98

[3] Ziemba P. Inter-criteria dependencies-based decision support in the sustainable wind energy management. Energies. 2019;**12**:749

[4] Ganji SRS, Rassafi AA, Kordani AA. Vehicle safety analysis based on a hybrid approach integrating DEMATEL, ANP and ER. KSCE Journal of Civil Engineering. 2018;**22**(11):4580-4592

[5] Brown RE, Gupta S, Christie RD, Venkata SS, Fletcher R. Distribution system reliability assessment using hierarchical Markov modeling. IEEE Transactions on Power Delivery. 1996;**11**(4):1929-1934

[6] Ataie E, Entezari-Maleki R, Rashidi L, Trivedi KS, Ardagna D, Movaghar A. Hierarchical stochastic models for performance, availability, and power consumption analysis of IaaS clouds. IEEE Transactions on Cloud Computing. 2019;**7**(4):1-18

[7] Nam T, Mavris DN. Multistage reliability-based design optimization and application to aircraft conceptual design. Journal of Aircraft, Georgia Institute of Technology, Atlanta, Georgia. 2018;**55**(5):1-15

[8] Paulson EJ, Starkey RP. Development of a multistage reliability-based design optimization method. Journal of Mechanical Design. 2013;**136**(1):1-8

[9] Wikström P, Terens LA, Kobi H. Reliability, availability, and maintainability of high-power variable-speed drive systems. IEEE Transactions on Industry Applications. 2000;**36**(1):231-241

[10] Bolvashenkov I, Kammermann J, Herzog H-G. Research on reliability and fault tolerance of multi-phase traction electric motors based on Markov models for multi-state systems. In: Proceedings of 23rd International IEEE Symposium on Power Electronics, Electrical Drives, Automation and Motion (SPEEDAM); 22th–24th June 2016; Anacapri, Italy. pp. 1-6

[11] Linkov I, Fox-Lent C, Read L, et al. Tiered approach to resilience assessment. Journal Risk Analysis. USA. 2018;**38**(4):1-9. DOI: 10.1111/risa.12991

[12] Woo S, O'Neal DL. Reliability design and case study of mechanical system like a hinge kit system in refrigerator subjected to repetitive stresses. Engineering Failure Analysis. 2019;**99**:319-329

[13] Xiao N, Huang N-Z, Li Y, He L, Jin T. Multiple failure modes analysis and weighted risk priority number evaluation in FMEA. Engineering Failure Analysis. 2011;**18**:1162-1170

[14] Ding Y, Lin Y, Peng R, Zuo MJ. Approximate reliability evaluation of large-scale multistate series-parallel systems. IEEE Transactions on Reliability. 2019;**68**(2):1-15

[15] Xuy X, Liz Z, Chen N. A hierarchical model for lithium-ion battery degradation prediction. IEEE Transactions on Reliability. 2016;**65**(1):310-325

[16] Abbas M, Vachtsevanos GJA. System-level approach to fault progression analysis in complex engineering systems. In: Proceedings of Annual Conference of the Prognostics and Health Management Society; September 27-October 1 2009; San Diego, CA. 2009. pp. 1-7

[17] Gomes JPP, Rodrigues LR, Galvão RKH, Yoneyama T. System level RUL estimation for multiple-component systems. In: Proceedings of Annual Conference of the Prognostics and Health Management Society; 14th–17th October 2013; New Orleans, LA, USA. 2013. pp. 1-9

[18] Rodrigues LR. Remaining useful life prediction for multiple-component systems based on a system-level performance indicator. IEEE/ASME Transactions on Mechatronics. 2018;**23**(1):1-10

[19] Bolvashenkov I, Kammermann J, Willerich S, Herzog H-G. Comparative study of reliability and fault tolerance of multi-phase permanent magnet synchronous motors for safety-critical drive trains. In: Proceedings of the International Conference on Renewable Energies and Power Quality (ICREPQ'16); 4th–6th May; Madrid, Spain. 2016. pp. 1-6

[20] Bolvashenkov I, Herzog H-G, Frenkel I, Khvatskin L, Lisnianski A. Safety-Critical Electrical Drives: Topologies, Reliability, Performance. Switzerland: Springer; 2018

[21] Bolvashenkov I, Kammermann J, Willerich S, Herzog H-G. Comparative study for the optimal choice of electric traction motors for a helicopter drive train. In: Proceedings of the 10th Conference on Sustainable Development of Energy, Water and Environment Systems (SDEWES'15); 27th Sept.–3rd Oct. 2015; Dubrovnik, Croatia. 2015. pp. 1-15

[22] Kammermann J, Bolvashenkov I, Herzog H-G. Reliability of induction machines: Statistics, tendencies, and perspectives. In: Proceedings of 26th IEEE International Symposium on Industrial Electronics (ISIE); 19th–21th June 2017; Edinburgh, UK. 2017. pp. 1843-1847

[23] Bolvashenkov I, Frenkel I, Kammermann J, Herzog HG. Comparison of the battery energy storage and fuel cell energy source for the safety-critical drives considering reliability and fault tolerance. In: Proceedings of IEEE International Conference on Information and Digital Technologies (IDT); 5th–7th July 2017; Žilina, Slovakia. pp. 63-70

[24] Bolvashenkov I, Kammermann J, Herzog H-G. Methodology for determining the transition probabilities for multi-state system markov models of fault tolerant electric vehicles. In: Proceedings of the Asian IEEE Conference on Energy, Power and Transportation Electrification; 25th–27th October 2016; Singapore. pp. 1-6

[25] Bolvashenkov I, Kammermann J, Herzog HG, Frenkel I. Operational availability and performance analysis of the multi-drive multi-motor electric propulsion system of an icebreaker gas tanker for arctic. In: Proceedings of IEEE 14th International Conference on Ecological Vehicles and Renewable Energies (EVER'19); 8th–10th Mai 2019; Monaco. 2019. pp. 1-6

[26] Bolvashenkov I, Kammermann J, Herzog H-G, Frenkel I, Ikar E, Khvatskin L. Investigation of reliability and fault tolerance of multiphase traction electric motor supplied with multi power source based on Lz-transform. In: Proceedings of IEEE International Conference on System Reliability and Safety (ICSRS'17); 20th–22th December 2017; Milano, Italy. 2017. pp. 303-309

[27] Bolvashenkov I, Herzog H-G. Use of stochastic models for operational efficiency analysis of multi power source traction drives. In: Proceedings of the Second IEEE International Symposium on Stochastic Models in Reliability Engineering, Life Science and Operations Management, (SMRLO); 15th–18th February 2016; Beer Sheva, Israel. pp. 124-130

[28] Frenkel I, Bolvashenkov I, Herzog H-G, Khvatskin L. Operational Sustainability Assessment of Multi Power Source Traction Drive. Mathematics Applied to Engineering. London, UK: Elsevier; 2017. pp. 191-203

[29] Bolvashenkov I, Kammermann J, Herzog H-G, Frenkel I. Fault tolerance assessment of multi-motor electrical drives with multi-phase traction motors based on LZ-transform. In: Proceedings of IEEE 14th International Conference on Ecological Vehicles and Renewable Energies (EVER'19); 8th–10th Mai 2019; Monaco. 2019. pp. 1-6

Chapter 3

Reliable Positioning and Journey Planning for Intelligent Transport Systems

Ahmed El-Mowafy, Nobuaki Kubo and Allison Kealy

Abstract

Safety and reliability of intelligent transport systems applications require positioning accuracy at the sub-meter level with availability and integrity above 99%. At present, no single positioning sensor can meet these requirements in particular in the urban environment. Possible sensors that can be used for this task are first reviewed. Next, a suggested integrated system of low-cost real-time kinematic (RTK) GNSS, inertial measurement units (IMU) and vehicle odometer is discussed. To ensure positioning integrity, a method for fault detection in GNSS observations and computation of the protection levels (PL) that bound the position errors at a pre-set risk probability of the integrated sensors are presented. A case study is performed for demonstration. Moreover, to save energy, reduce pollution, and to improve the economy of the trip, proper journey planning is required. A new approach is introduced using 3D city models to predict the route with the best positioning integrity, availability and precision for route selection among different possible routes. The practical demonstration shows that effectiveness of this method. Finally, the potential of using the next generation SBAS for ITS applications was tested using kinematic tests carried out in various environments characterized by different levels of sky-visibility that may affect observations from GNSS.

Keywords: intelligent transport systems, positioning, GNSS, IMU, odometer, integrity monitoring, SBAS, prediction

1. Introduction

Intelligent transport systems (ITS) require reliable, continuous, accurate and cost-effective vehicle positioning in real time and in different weather and work conditions with for lane identification and control of vehicles. Normally, the width of the lane is about 2.8–3.5 m and that of the car is about 2 m. Therefore, better than 1 m horizontal accuracy is required to recognize the lane and 0.1 m horizontal accuracy is needed to control the vehicle [1]. The work environment of ITS is also very dynamic, changing between open sky, semi-urban to urban where vehicles may be surrounded by other vehicles and travel in overpass, tunnels, etc. In such dynamic environment, and due to limitations in the capabilities, and performance of available positioning sensors, such as global navigation satellite systems (GNSS), inertial measuring units (IMU), odometers, cameras and radar, it is hard to depend solely on one system, but rather on integrated set of sensors. For example, some ITS applications rely on light detection and ranging with the help of cameras and radar. However, cameras sometimes cannot recognize the lane since the white line sometimes disappeared, and cameras are less reliable during foggy weather and at night. Similarly, while GNSS provides a primary positioning system, its signals can easily be obstructed in the urban environment. Thus, GNSS needs to be integrated with other sensors such as inertial measurement units (IMUs) to bridge positioning during GNSS positioning outages [2, 3]. To help the on-board positioning systems, methods such as cooperative positioning were proposed [4]. When selecting sensors, one should note that their cost varies much according to their quality; yet, the car industry can only afford a combination of sensors with a total cost that does not exceed a few hundreds of dollars. Therefore, low cost sensors, which may have limited capabilities, are used.

For ITS safety applications, not only accuracy is needed, but also integrity and reliability. Positioning integrity is a key performance parameter, where the system should be able to detect and exclude measurement faults, bound measurement errors, and trigger an alarm in the event that unreliable positioning performance is suspected. In addition, when no fault is detected, a protection level (PL) should be computed to bound the true position error at a certain probability of risk [5]. Integrity monitoring (IM) is currently being applied in aviation using an advanced receiver autonomous integrity monitoring (ARAIM) approach, which relies on the use of multi-frequency and multi-constellation phase-smoothed pseudo-range observations [6–8]. However, limited research has been done for applications such as ITS which require precise positioning that utilizes carrier-phase observations. Some examples are given in [9] when positioning using relative positioning, in [10] using precise point positioning (PPP) method, and in [11] using Real-Time Kinematic (RTK) method. At the moment, integrity requirements in vehicular transport have not been defined yet, but the demand for higher levels of automation in an increasing number of applications is pushing the relevant authorities to urgently fill this gap.

Up to now, most IM proposed methods focused on applying ARAIM in aviation and only employing GNSS measurements. Such integrity monitoring for transport applications has been addressed in a few studies, for instance in [12–14], where the focus was primarily on the use of single-frequency code observations. However, for ITS, GNSS cannot be used solely, and hence new IM methods are needed when integrating GNSS with other sensors [15]. In this article, the PLs that bound the position error are presented for continuous positioning of vehicular applications by integrating, in a simple fusion, loosely coupled algorithm low-cost RTK GNSS using code and phase observations supplemented by Doppler measurements, combined with low-cost IMU and vehicle odometer data. For GNSS we restrict our focus to horizontal positioning for the along-track and cross-track positioning of the vehicles, which are of interest for lane identification and collision warning.

For journey planning, to ensure reliability, and to save energy and reduce trip time, prediction of GNSS positioning integrity and precision are presented using 3D city models [16, 17]. The 3D city model also helps in identifying non-line of sight (NLOS) GNSS signals, which is a source of error. In addition, while RTK or network RTK [18] requires data from reference stations, the use of satellite based augmentation systems (SBAS) [19], does not have this requirement where sub-m accuracy in a stand-alone mode can be obtained. Therefore, results from testing the new generation of SBAS of the Australian test bed applied for ITS in various work environments are presented. This SBAS comprises the traditional L1 legacy SBAS signals for GPS only, the new dual-frequency multi-constellation (DFMC) SBAS and SBAS-based PPP using measurements from GPS and Galileo measurements. The following sections describe these methods, and their performance is demonstrated through tests representing ITS applications.

2. Positioning in ITS

2.1 GNSS and IMU as the main positioning sensors

There is a range of GNSS methods that can be used for transportation applications. Their features and accuracy are summarized in **Table 1**. The single point positioning (SPP) and Differential GPS (DGPS) use only one receiver, and employ single-frequency undifferenced code observations for the former and with corrections of satellite-related errors (satellite orbit and clocks corrections) in the latter, making them affordable and widely used for vehicle navigation. However, both approaches provide several meters of positioning error, and thus they are not suitable for ITS. With the sub-m requirement of ITS, only three methods can be used, namely real-time kinematic (RTK) [20] or network RTK (NRTK) [18], precise point positioning (PPP) [21, 22], and the next generation SBAS [19]. The advent of low-cost dual-frequency multi-constellation GNSS, at the level of a few hundreds of dollars, allow their use in advanced vehicle positioning. Their performance has recently been remarkably improved, at a few cm accuracy.

For the IMU (also known as inertial navigation system (INS), typically after obtaining a navigation solution), the strategic grade type provides the best performance, but at a high cost and thus is not suitable for vehicle applications. However, small, robust, and low-cost inertial sensors, e.g. the micro electrical mechanical sensors (MEMS) IMUs [23], have been available in the market for several years, which can be used in vehicle navigation. They, however, suffer from the rapid growth of their biases. The solutions obtained from GNSS and IMU complement each other, as they have different characteristics, summarized in **Table 2**. GPS solution aid IMU by resetting the accumulation of its bias. On the other hand, IMU can extrapolate solutions at a higher rate and can cover positioning during short GNSS outages. IMU additionally provides the attitude (orientation) that can also be

Features	**SPP**	**DGPS**	**SBAS**	**RTK (Sing Ref. and NRTK)**	**PPP**
No. of receivers	one	one	one	2 (RTK) 1 (NRTK)	one
No. of Observation Frequencies	one (e.g. L1)	one (e.g. L1)	one (e.g. L1) 2 DFMC (L1&L5)	2 (e.g. L1&L2 or L1&L5)	2 (e.g. L1&L2 or L1&L5)
Need for data or corrections	Autonomous	Reference station corrections	Orbit + clock corr. (+iono for L1)	Reference station data	Orbit + clock corrections
Main observations	Code obs.	Code obs.	Code obs.	Code + phase obs.	Code + phase obs.
Accuracy	1–6 m	Sub-m to a few-m	Sub-m to 1.5 m	Few cm	Dm (float), cms (fixed)
Solution type	Snap shot	Snap shot	Snap shot	Kalman or sequential LS	Kalman filter or seq. LS
Main issues	Low accuracy - noisy	Low accuracy - noisy	Not suitable for urban	Ambiguity fixing	Long time for Convergence
Suitable for ITS	No	No	In open sky only	Yes	Yes

Table 1.
Features of GNSS methods used for positioning in transportation.

used in estimating the positioning errors along the vehicle direction of motion, which is needed for a more representative integrity monitoring as will be explained later, and in applications such as collision alert.

2.2 Simple integration of low-cost GNSS, IMU and odometer

In this article, low-cost systems that are suitable for vehicle applications are considered. Two approaches can be applied to control the growth of heading bias of the MEMS IMU. At the start, or when the vehicle stops, e.g. at red traffic lights, the zero velocity update (ZUPT) is applied. When GNSS data is available, it is used to reset the heading bias of the MEMS IMU. The GNSS position and velocity are coupled with the IMU output using Kalman Filter in loosely- or tightly-coupled schemes. While the tightly coupled integration is beneficial in the case when GNSS cannot estimate the position, e.g. due to a low number of visible satellites, the IMU data can be used to slightly predict the pseudo-range observations; however, it is impossible to predict the carrier phase observations at the level of ambiguity fixed solution. Therefore, for the low-cost RTK/IMU systems, no practical difference exists between using loosely and tightly coupled integration.

In RTK a minimum of five satellites should be observed. When observing four satellites, e.g. in a semi-urban environment, a simple approach can be applied for positioning using the low cost systems. GNSS Doppler velocities can be used to compute initial values of IMU heading and to calibrate it at short intervals to control the growth of its bias [24]. The computed heading from Doppler measurements at time t (denoted as θ_t) is calculated from the average $\theta_t = tan^{-1}\left(\frac{V_{E_t}}{V_{N_t}}\right)$, where V_{E_t} and V_{N_t} are the Doppler-based velocity components in the local-level frame. Thus, the accuracy of heading obtained from GNSS depends on the velocity measurements and partly on the dilution of precision (DOP), which is an indicator of the number and geometry of observed satellites (their distribution in the sky). When the speed of the vehicle is low, the heading from GNSS is not reliable because the computed GNSS velocity would be noisy in the order of a few centimeters per second. In addition, the sampling rate of GNSS is less than that of the heading rate of the MEMS IMU; therefore, when the road suddenly bends, the obtained heading could deviate several degrees from the actual orientation of the vehicle. Consequently, several conditions were set to use GNSS for calibrating the IMU heading error to below two degrees. These conditions include VSS > 0.5 m/s; $|V_{GNSS} - V_{SS}|$ < 0.5 m/s; VSS > 0.5 m/s, where V_{GNSS} and V_{ss} are the velocities estimated by the GNSS and vehicle's odometer, respectively. Large errors of the heading obtained from the GNSS should not be used. Therefore, the heading from the GNSS must

GNSS	IMU
Absolute positioning	Relative positioning
Good accuracy in long term	Good accuracy in short term
Attitude available—need multiple units	Attitude available—single unit
Low sampling rate (1–10 Hz)	High sampling rate (e.g. 100 Hz)
Subject to signal blockage	Not related to surrounding environment
Low-cost can provide good accuracy	Low-cost provides poor accuracy
Biases are stable	Biases grow with time

Table 2.
Characteristics of GNSS and IMU.

always be checked, with a threshold in the order of 2°, using both the best estimated heading in the previous epoch and the heading rate of the IMU, as the heading rate obtained from the IMU in a short period is accurate.

If RTK is unavailable, the positions can be estimated by integrating the speed estimated from the vehicle odometer with the heading of the MEMS IMU. The time increments of position components in North and East (ΔE, ΔN) is computed, such that $\Delta E = V_{ss} \times \sin(\theta) \times \Delta t$ and $\Delta N = V_{ss} \times \cos(\theta) \times \Delta t$, where θ is the heading estimated by the IMU and Δt is the time increment. The velocity and azimuth considered are the mean values during the time increment. The odometer equipped in the vehicle is used to obtain the distance information of the car. Normally, a velocity pulse generation device counts the number of pulses per rotation of the wheel. In this study, we use the speed pulse obtained from the POSLVX system, which is a wheel-mounted rotary shaft encoder that accurately measures the linear distance covered by the vehicle [25]. The two methods, Doppler calibrated IMU and Odometer+IMU, can only estimate temporal position changes, and hence, their positioning errors accumulate with time, in particular, the heading bias of the IMU. Therefore, they should only be restricted to bridging short breaks in RTK as will be discussed later by an example.

2.3 SBAS for ITS

The positioning accuracy of the traditional SPP method can be improved by using orbital and clock corrections from the satellite based augmentation systems (SBAS). SBAS can provide meter-level accuracy in a stand-alone mode without the need for relative positioning with a nearby base station that is required in RTK. However, traditional SBAS systems, such as the United States WAAS system or the European EGNOS system augment only L1 single frequency measurements. Therefore, the ionosphere delays need to be processed and delivered to the users. This makes single-frequency SBAS sensitive to the distribution of the ground network used to compute the ionosphere corrections, have a limited coverage area, and are less precise during rapid fluctuations of the ionosphere.

The second-generation SBAS, such as that implemented in the under-development Australia SBAS, includes in addition to the traditional L1 legacy SBAS signals, dual-frequency multi-constellation (DFMC) SBAS signals that are transmitted over L5 for GPS L1/L5 and Galileo E1/E5a signals. Hence, the user can apply ionosphere-free combination without the need for ionosphere corrections. This allows the user to work anywhere within the footprint of the SBAS satellite and is not sensitive to ionosphere fluctuations. Moreover, the second generation SBAS includes precise orbits and satellite clock corrections to enable precise point positioning (PPP) service with float-ambiguity solution type. This can provide dm level accuracy, which is suitable for ITS. Furthermore, the multi-constellation scenario will increase the number of satellites and thus provides users with a better measurement geometry. Results from testing the new generation of SBAS when applied for ITS in various work environments are presented in the testing section.

2.4 Cooperative positioning

The recent developments in vehicular ad-hoc networks (VANETs) and dedicated short range communication (DSRC) support the principle of cooperative positioning (CP) through wireless connectivity. CP has been proposed to share positioning information obtained from each individual vehicle, connected to the system, to increase the vehicle awareness of the surrounding vehicles, predict potential incidents, threats, and hazards on the road with an increased time horizon and awareness distance that is beyond what in-vehicle technologies (radars or cameras) and the driver can visualize [26].

In the vehicle-to-vehicle (V2V) and vehicle-to-infrastructure (V2I) communications, vehicles send messages to each other or to infrastructure. These messages include their temporary ID, location, speed, heading, lateral and longitudinal acceleration, brake system status, and vehicle size [27]. Sharing this data in the V2V communication can provide warnings to the drivers in poor vision scenarios during the rear end or intersection collision and lane change. Examples of V2I benefits include awareness of unsafe conditions on the road, including fog, ice, and Eco-approach and departure at signalized intersections.

The positions of all nodes in the in the VANET network can be determined by integrating this information in either a centralized or decentralized algorithm. An alternative concept relies either on the availability of ranging information to other vehicles using V2V communication or the availability of ranging information to DSRC roadside units (RSUs) using V2I [28]. Other CP methods leverage the communications signal as a ranging signal. For instance, methods such as signal strength-based ranging, time-based ranging including time of arrival (TOA) and time difference of arrival (TDOA) methods have been implemented however, these produce errors at several meters, which makes them not suitable for ITS. On the other hand, Non Ranging-Based techniques do not rely on time or signal strength ranging techniques, and thus, have fewer errors, but they are expensive since it requires RSUs, installed at each intersection, storing information about the road geometry. Several approaches were proposed to enable CP in the framework of VANET, while each method has merit, it has also limitations. This idea of ranging between nodes in the network has led to sensors such as ultra wide band and radar being deployed as part of the multi-sensor suite to overcome some of the limitations associated with ranging from DSRC signals.

Based on this discussion of CP, as depicted in **Figure 1**, two distinct subsets of navigation systems are used to define a CP architecture. The first is termed as local level, where each vehicle takes its own measurements and would be able to provide its own position estimate, independent of other vehicles. This system typically consists of GNSS and IMU. The second is termed as network level, where vehicles would share information among each other to form a network, using DSRC and UWB, to provide a more robust position estimate in GNSS denied environments. The advantages of the network approach over the local level are: the additional inter-vehicle measurements provide greater measurement redundancy and consequently improve the precision and robustness of the solution. In sharing measurements, vehicles with insufficient measurements to determine a local level solution, are still able to determine their position, thereby improving the availability of positioning solutions across the whole network. A network approach is also scalable to a large number of nodes,

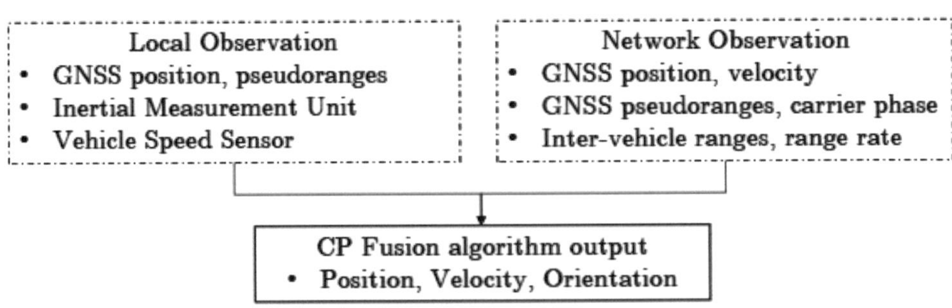

Figure 1.
Local and network level observations.

using a decentralized processing approach, a large network can share information in a way that maintains the optimal estimate for each node, whilst balancing the computational overheads. This allows this approach more computationally efficient compared to centralized processing at each node in the local level.

3. Route selection

For ITS, reliable real-time positioning should be maintained all the time. In addition, for the driver convenience and to save energy and reduce pollution by reducing journey time and improving the economy of a trip, proper journey planning is required by the computer on board the ITS vehicle before the start of the Journey. The route with best positioning performance in terms of integrity availability and precision, in addition to other parameters such as trip time and distance, is selected among different possible routes. Prediction of these parameters can be best performed using 3D city models when using GNSS, where a minimum of five satellites need to be observed to enable GNSS-RTK positioning. The method is also suitable to determine locations within the city, at different times of the day that GNSS would not be suitable and other sensors should be used.

The general procedure to predict the satellites that are in view and the existence of LOS (LINE-OF-SIGHT) between a GNSS satellite and GNSS receiver using a 3D city model for each location along candidate routes comprises the following four steps:

a. *Identify the receiver position*: The vehicle position is first approximately estimated at each epoch according to time and location of the start of the journey, speed of the vehicle (taken as the road speed), and the centerline of the lane that has the worst satellite geometry, typically the nearest to the nearby buildings. A change of position of ~20 m/s is expected, therefore, the LOS checking should be performed at short time intervals, e.g. 2–10 Hz.

b. *Estimate satellite positions*: The satellite positions are predicted using satellite navigation file, according to the expected time of processing. Due to the slow change of satellite geometry, an actual delay or advance of a few minutes between the actual and prediction time would not make a significant difference in the satellite geometry, and hence in the prediction result.

c. *Establishment of buildings and terrain data*: The location, size, height and orientation of each building along the route are constructed from 3D polygon data. Buildings and terrain data are prepared separately and later are combined. **Figure 2** shows an example of 3D models that were created for our research. More details will be given in the testing section.

d. *Assessing the existence of LOS between satellites and the receiver*: The presence of LOS between possible satellites in view from the almanac and the receiver can be determined by checking whether the direct line connecting the satellite and the receiver intersects with the building. By modeling the building as triangular meshes, it is possible to determine the presence of an intersection as shown in **Figure 3** using the following method presented in [29]. In this figure, S indicates the satellite position and R represents the receiver's location. B_1, B_2 and B_3 are the vertices of a triangular face of the building meshed surface facing the receiver. Let s, r, b_1, b_2 and b_3 be the corresponding position vectors in an arbitrary reference frame, and u_{ij} be the unit vector from B_i to B_j. Likewise, let point P be the intersection between the vector

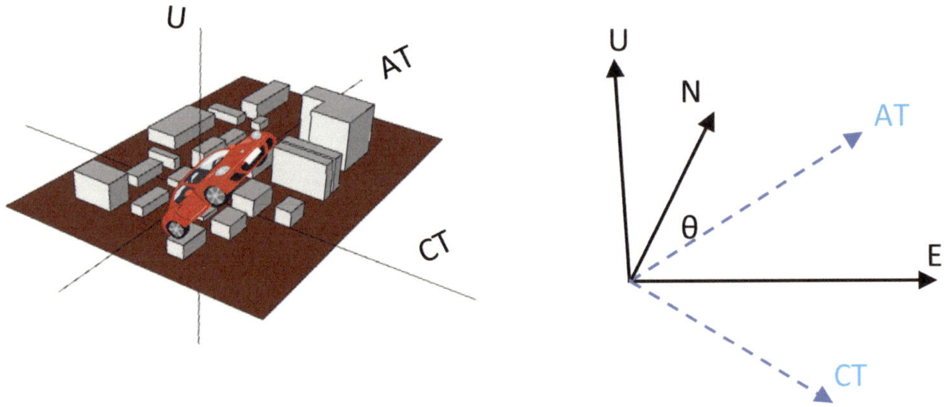

Figure 2.
Example of buildings and terrain data of 3D city model (left), showing the AT-CT and N-E (right). θ is the azimuth.

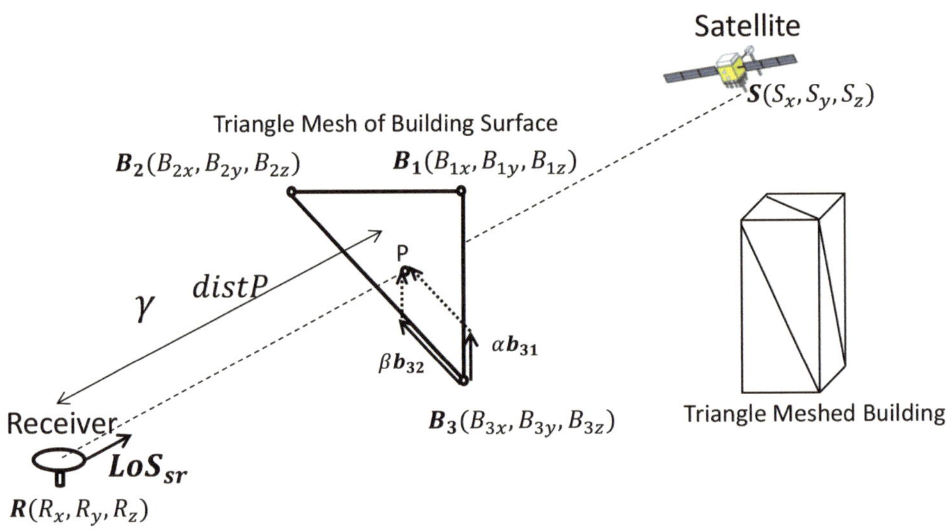

Figure 3.
Prediction of the satellite LOS.

between the receiver and the satellite, which unit direction vector is indicated by u_{LOS}, and the triangular mesh, and p is the corresponding position vector. In the vector space we have:

$$p = r + \gamma\, u_{LOS} \text{ and } p = b_1 + \alpha\, u_{13} + \beta\, u_{12} \quad (1)$$

where γ is the distance between R and P. α and β are the coordinates of P in a local frame that includes the sides of the mesh triangle as principle axes. The system in (1) can be solved for γ, α, and β. Thus, the conditions for P to be included in the triangular mesh, and therefore No LOS (NLOS) are:

$$\gamma \geq 0; 1 \geq \frac{\alpha}{|b_1 - b_3|} \geq 0;$$
$$1 \geq \frac{\beta}{|b_1 - b_2|} \geq 0; 1 \geq \frac{\alpha}{|b_1 - b_3|} + \frac{\beta}{|b_1 - b_2|} \geq 0 \quad (2)$$

Otherwise, the satellite position is considered in LOS. The geometry of the visible satellites is then examined to check sufficient positioning precision can be obtained, and check integrity as will be discussed in the next section.

4. Integrity monitoring

Integrity monitoring includes the ability of the system to detect and exclude faulty observations (known as FDE) and to alarm the user if a protection level (PL) that bounds the true position error at a specific risk probability is less than an alarm limit (AL). Integrity monitoring is considered available when PL < AL. Both tasks, i.e. FDE and computation of PL, are addressed in the following sub-sections.

4.1 Fault detection

In general, the equation of the fault-free observations can be expressed as:

$$y = G\,x + \varepsilon \tag{3}$$

where y is the measurement vector, computed as the difference between the observations and their estimated values from the approximate user and satellite positions. The null hypothesis is expressed as H_0: E{y} = $G\,x$ with D{y} = Q_y, representing the covariance matrix of the observations, where E{} and D{} denote the expectation and dispersion operators, respectively. The unknown vector x is the difference between the final and the approximate vehicle's computed positions. ε is the observation error, assumed noise in the fault-free case with zero mean and Gaussian distribution. The G matrix for RTK is the direction cosine matrix.

For the IMU + odometer, the observations are the Easting and Northing velocity components computed as $V_E = V_{ss} \times \sin(\theta)$, and $V_N = V_{ss} \times \cos(\theta)$. These velocities are integrated in time to provide the time changes in position in Easting and Northing directions. The observations are considered in this case as the mean values of the IMU heading (θ) and the odometer speed V_{ss}, for instance between the epochs *t-1* and *t*. Thus, the G matrix is expressed as [25]:

$$G_t = \begin{bmatrix} \frac{\partial\theta}{\partial\Delta E} & \frac{\partial\theta}{\partial\Delta N} \\ \frac{\partial v}{\partial\Delta E} & \frac{\partial v}{\partial\Delta N} \end{bmatrix}_t = \begin{bmatrix} \frac{\Delta N}{\Delta E^2 + \Delta N^2} & \frac{-\Delta E}{\Delta E^2 + \Delta N^2} \\ \frac{\Delta E}{\Delta t \times \sqrt{\Delta E^2 + \Delta N^2}} & \frac{\Delta N}{\Delta t \times \sqrt{\Delta E^2 + \Delta N^2}} \end{bmatrix}_t \tag{4}$$

Using least squares for fault detection, the solution in (E-N-U) frame reads:

$$\hat{x} = R\left(G^T Q_y^{-1} G\right)^{-1} G^T Q_y^{-1} y = S\,y \tag{5}$$

where $S = R\left(G^T Q_y^{-1} G\right)^{-1} G^T Q_y^{-1}$ is the pseudo inverse, which maps the observations onto the unknowns. In RTK, R is the rotation matrix from the Cartesian frame, in which the GNSS satellite positions are expressed, to the E-N-U frame. When using IMU + odometer measurements, R is the identity matrix. To identify which observations are faulty, the solution separation method can be applied [6]. This is performed by computing a position solution unaffected by the fault, by excluding the suspected observations. An error bound around this solution is computed, and the difference between the position solution from all observations and the fault tolerant position is accounted for. For each potential fault mode i, which may comprise one or more faulty observations, an analogous S_i matrix is computed by excluding the suspected observations, such that:

$$\hat{x}_i = S_i\, y \tag{6}$$

The discrepancy in the positional vector $|\hat{x} - \hat{x}_i|$ forms the base for checking the presence of observation faults, where in case of faulty measurements in mode i, the difference between the two solutions $\hat{x}$ and $\hat{x}_i$ will be significant. The standard deviations of this difference ($\sigma_{dE_i}, \sigma_{dN_i}, \sigma_{dU_i}$) are next computed as:

$$\sigma_q = \sqrt{a_k^T (S_i - S) Q_y (S_i - S)^T\, a_k} \tag{7}$$

k = 1, 2, 3 for dE_i, dN_i, dU_i, with $a_1^T = [1,0,0], a_2^T = [0,1,0]$, and $a_3^T = [0,0,1]$.

For ITS applications, where only horizontal positioning is considered, it is more convenient to conduct testing for the along-track (AT) and cross-track (CT) position directions [12, 30]. Assuming $\Delta\hat{x}_i = |\hat{x} - \hat{x}_i|$ has a zero-mean Gaussian distribution in the fault-free mode; and considering its components for the AT and CT directions, defined as $\Delta\hat{x}_{i_{AT}}$ and $\Delta\hat{x}_{i_{CT}}$. Then the normalized discrepancies, i.e. $\frac{|\Delta\hat{\mathrm{x}}_{iAT}|}{\sigma_{\Delta\hat{x}_{iAT}}}$ and $\frac{|\Delta\hat{\mathrm{x}}_{iCT}|}{\sigma_{\Delta\hat{x}_{iCT}}}$ will also have a zero-mean Gaussian distribution, and will be used as the test statistic, where $\sigma_{\Delta\hat{x}iAT}$ and $\sigma_{\Delta\hat{x}iCT}$ are the stds of $\Delta\hat{x}_{i_{AT}}$ and $\Delta\hat{x}_{i_{CT}}$ respectively. Therefore, when examining m possible fault modes, for i = 1 to m, setting a threshold of the standard normal distribution at a selected significance level, i.e. $N_{\frac{\alpha}{2\times 2m}}(0,1)$, a fault is suspected when:

$$H_{aiAT} : \frac{|\Delta\hat{\mathrm{x}}_{iAT}|}{\sigma_{\Delta\hat{x}_{iAT}}} \geq N_{\frac{\alpha}{2\times 2m}}(0,1) \text{ and } H_{aiCT} : \frac{|\Delta\hat{\mathrm{x}}_{iCT}|}{\sigma_{\Delta\hat{x}_{iCT}}} \geq N_{\frac{\alpha}{2\times 2m}}(0,1) \tag{8}$$

When the direction of the vehicle is not well defined, or when this direction rapidly changes, for instance during rapid turns, a conservative approach is to perform the FDE test in the direction of the maximum error. To this end, the maximum-minimum region defined along the semi-major and semi-minor axes of a horizontal confidence error ellipse can be tested. These directions are defined in the Eigen space by the orientation of the first and second Eigen vectors. The semi-major axis of the error ellipse, which represents the max std. $\sigma_{\Delta x_{imax}}$ equals $\sqrt{\xi 1_i}$, where $\xi 1_i$ is the first Eigenvalue. Similarly, the semi-minor axis of this error ellipse $\sigma_{\Delta x_{imin}}$ is calculated as $\sqrt{\xi 2_i}$, where $\xi 2_i$ represents the second Eigenvalue of $Q_{\Delta\hat{x}_i}$., the 2D variance matrix of $\Delta\hat{x}_i$ computed by applying the propagation law. Thus, the null hypothesis is rejected suspecting a fault in mode i when:

$$\frac{\left(\Delta\hat{\mathrm{x}}_i^T \cdot \vec{E}1_i\right)^2}{\sigma^2_{\Delta x_{imax}}} + \frac{\left(\Delta\hat{\mathrm{x}}_i^T \cdot \vec{E}2_i\right)^2}{\sigma^2_{\Delta x_{imin}}} \geq \chi^2_{\frac{\alpha}{m}}(df, 0) \tag{9}$$

where $\vec{E}1_i$ and $\vec{E}2_i$ denote the first and second Eigenvector of $Q_{\Delta\hat{x}_i}$.

4.2 Computation of the protection levels

In our work, the *PLs* are modeled on the basis of the multi-hypothesis solution-separation method [6]. In RTK, with $df > 0$, a position error bound is computed for each possible fault mode i that might be misdetected. The *PLs* can be computed by solving the equations [12, 30]:

$$2\,\psi\left(\frac{PL_{AT}-b_{oAT}}{\sigma_{\hat{x}oAT}}\right)+\sum_{i=1}^{N_a}P_i\psi\left(\frac{PL_{AT}-K_{fa}\sigma_{\Delta\hat{x}iAT}-b_{iAT}}{\sigma_{\hat{x}iAT}}\right)=\frac{1}{2}(P_{HMIH}) \quad (10)$$

$$2\,\psi\left(\frac{PL_{cT}-b_{ocT}}{\sigma_{\hat{x}ocT}}\right)+\sum_{i=1}^{N_a}P_i\psi\left(\frac{PL_{cT}-K_{fa}\sigma_{\Delta\hat{x}icT}-b_{icT}}{\sigma_{\hat{x}icT}}\right)=\frac{1}{2}(P_{HMIH}) \quad (11)$$

The components (b_{oAT} and b_{oCT}) and (b_{iAT} and b_{iCT}) are projected in the position space using S and S_i from b_o and b_i, which denote the sum of the maximum nominal biases in the observations under the fault-free and fault hypotheses, respectively. $\sigma_{\hat{x}oAT}$ and $\sigma_{\hat{x}iAT}$ are the stds for the AT position solution and $\sigma_{\hat{x}oCT}$ and $\sigma_{\hat{x}iCT}$ are for CT. Similarly, $\sigma_{\Delta\hat{x}iAT}$ and $\sigma_{\Delta\hat{x}iCT}$ are the stds of $\Delta\hat{x}_{i_{AT}}$ and $\Delta\hat{x}_{i_{CT}}$ respectively. The absolute value of the bias is considered to bound the worst case scenario and to ensure that the continuity requirement is met. $\psi(\ )$ is the tail probability of the cumulative distribution function of a Gaussian distribution. P_i is the a-priori probability of fault in the observations in the examined fault mode i, assuming the same probability for all observations from one system, which differ among systems. Since no standards are available yet for IM in ITS, P_{HMIH} of 10^{-5} is assumed. The K_{fa} becomes $\psi^{-1}\left(\frac{\alpha}{2m}\right)$, where we assume α = 1% in this article.

It is assumed here that the observation and the position errors follow a Gaussian distribution. In the open sky environment, this is valid, but in the urban environment, and due to multipath, the distribution could be biased and it could have multiple peaks. Additionally, in ITS and due to motion of the vehicle and the dynamic change of the nearby structures, causing multipath, multipath tends to randomize. One approach here is to deweight the observations that may experience large multipath, such that their contribution in the solution is minimized. Such multipath modeling is based on satellite elevation and azimuth, and the use of 3D city models to describe the geometry of the surrounding structures. Moreover, an overbounding Gaussian distribution with large stds [31] can be used. However, this would lead to large PLs, a loss in integrity availability. **Figure 4** shows a flowchart of the positioning and integrity monitoring process, which includes the FDE and computation of the PLs.

For the case of integrating the heading from MEMS IMU and speed estimated by the odometer, where df = 0, the PLs are expressed as:

$$PL_{AT,i}=K_{md,i}\sigma_{AT,i}+\sqrt{\left(\cos\theta a_1^T S\begin{bmatrix}b_{\theta_{IMU}}\\ b_v\end{bmatrix}\right)^2+\left(\sin\theta a_2^T S\begin{bmatrix}b_{\theta_{IMU}}\\ b_v\end{bmatrix}\right)^2}$$

$$PL_{CT,i}=K_{md,i}\sigma_{CT,i}+\sqrt{\left(\sin\theta a_1^T S\begin{bmatrix}b_{\theta_{IMU}}\\ b_v\end{bmatrix}\right)^2+\left(\cos\theta a_2^T S\begin{bmatrix}b_{\theta_{IMU}}\\ b_v\end{bmatrix}\right)^2} \quad (12)$$

where $b_{\theta_{IMU}}$ is a scalar representing possible unaccounted for growth in IMU heading bias between the bias resetting, using, for instance, GNSS heading. It is assumed here that this bias increases linearly with time, such that $b_{\theta_{IMU}} = b_{\theta_o}+\Delta b\times\Delta t$, where b_{θ_o} is the initial bias, Δb is the bias drift with time, and Δt is the time difference between current epoch and the resetting epoch. b_v denotes the bias due to velocity measured by the odometer. $K_{md,i}$ is the inverse of the complement of the right-side standard normal cumulative distribution function (i.e. $K_{md,i}=\psi^{-1}(\beta)$) to satisfy the misdetection probability β, which can be preset. The final protection levels PL_{AT} and PL_{CT} are considered as the max$\{PL_{AT,i}\}$ and max $\{PL_{CT,i}\}$.

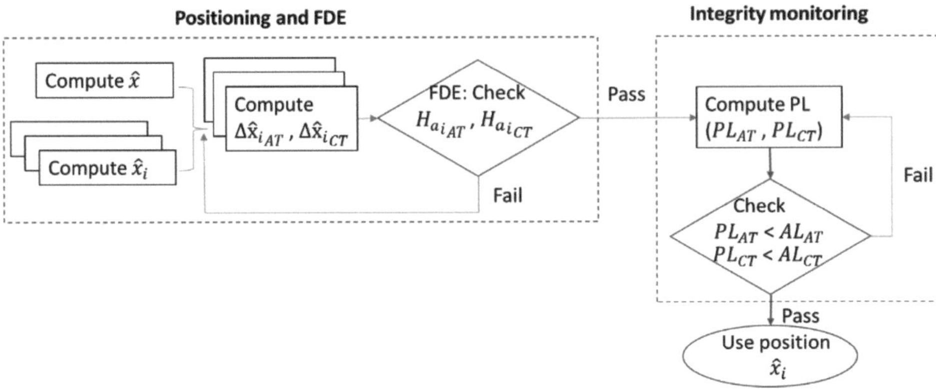

Figure 4.
Flowchart of the positioning, FDE and integrity monitoring process.

Since in the IMU + odometer case only time-changes of positions are measured, integrity risk has to additionally consider the accumulation of errors. The covariance matrix of the unknown coordinates (Q_{EN_t}) at time t can be expressed as:

$$Q_{EN_t} = G\,Q_{obs}G^T + Q_{EN_{t-1}} \tag{13}$$

To bound the development of the accumulated error, this positioning approach needs to be reinitialized at short time intervals and the Q_{EN_k} is reset with each reinitialization. Such an approach will lead to a sawtooth-like pattern for the PL to adapt to the growth-reset error pattern.

5. Testing

In this section test results of an integrated GNSS/IMU/odometer system proposed for ITS are first presented and discussed. Afterwards, the next generation SBAS is tested when being implemented within an ITS setting.

5.1. Test description

A kinematic test was conducted using a small vehicle fitted with a low-cost RTK, MEMS IMU and odometer. The test is performed in semi-urban and urban environments in Tokyo, Japan. The road has 2–3 lanes on each side of the road, where several high-rise buildings were present. In addition, several overpasses, pedestrian bridges, and a river bridge were also present. The test trajectory is shown in **Figure 5**. The use of multi-system GNSS measurements is essential to observe enough number of satellites for RTK positioning and to resolve the ambiguities as quickly as possible to maintain reliability in this environment as the number of observed satellites changes frequently between 5 and 22. Therefore, the RTK system used GPS, GLONASS, QZSS and BeiDou dual-frequency observations with 10 Hz sampling interval. The RTK-GNSS was supported by Doppler frequency observations [32]. The Doppler-aided RTK-GNSS usually improves the fix rate by about 10–15% and provide the same reliability [33]. The positioning error (PE) in the RTK mode was estimated as the difference between the RTK-computed positions and those determined from post-mission kinematic processing (PPK) of the same data collected by the receiver but using independent software. The vehicle receiver operates within a few kilometers from a reference receiver occupying a known

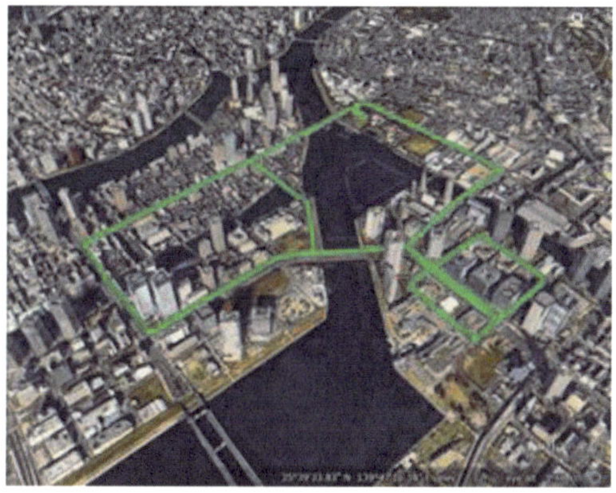

Figure 5.
Test trajectory in Tokyo.

point such that the spatially correlated errors; i.e. the broadcast orbital error and the atmospheric delays - ionosphere and troposphere, are canceled by double differencing the observations between the rover and the reference receiver, leading to ±5 cm accuracy [25].

A Bosch-consumer grade MEMS IMU was used in the test. The heading error of this IMU ranged from −2 to 5°, which can accumulate to 10° after 30 min without calibration [25, 34]. The raw 3-axis angular rate, 3-axis acceleration, and pressure are provided. The vertical position of the GNSS is integrated with the change in the vertical position deduced from the barometric sensor. If the velocity vector estimated by the GNSS is not available, the velocity vector estimated by the final integrated heading and speed sensor is used instead. If the absolute difference between these two velocity vectors is more than 0.25 m/s, we rely on the velocity vector estimated by the final integrated heading and odometer. The sampling rate of IMU and odometer was 100 Hz. For the odometer, the standard deviation of the computed speed is estimated as 5 cm/s, and for the speed determined from GNSS-Doppler measurements, it is 10 cm/s. The positioning errors when using IMU + odometer were computed by differencing their positions with the output from a POS/LV system (developed by Applanix Inc.), which was mounted on the vehicle and has a nominal positioning accuracy of approximately 20 cm.

The test was also used to demonstrate the performance of the prediction approach for selection of the best route based on best integrity availability and trip characteristics, such as distance and time of travel. Results of the predicted satellite positions and their geometry applying the 3D city model are compared with the real observed satellites obtained for the same route and the same period. The 3D city model is built using a GEOSPACE 3D solution numerical surface with polygon representation (http://www.ntt-geospace.co.jp/geospace/3d.html). The model includes land, roads, bridges, buildings, and vegetation. The 3D buildings are created by adding height information to the GEOSPACE digital 2D map, where the heights are measured from GEOSPACE aerial photographs (orthoimage). The accuracy of the models employed in this test is 2–3 m.

5.2 Route selection results

Figure 6(left) shows the difference between the number of satellites in view determined by the prediction algorithm and the number of the actually observed satellites along the test route. The difference in their satellite geometry, expressed by Position Dilution of Precision (PDOP), is illustrated in **Figure 6(right)**. The

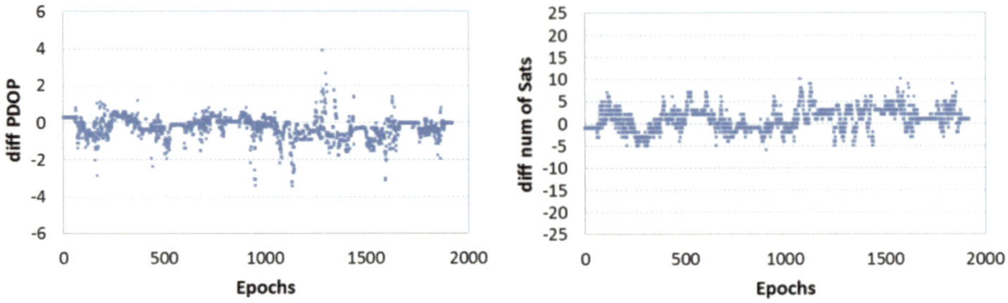

Figure 6.
Difference between using observed and predicted satellites in terms of: (left) number of satellites and (right) geometry expressed by PDOP.

results show that the average difference between the number of observed and predicted satellites in view is two satellites, ranging between ±5 satellites. The mean value of the absolute difference in PDOP is 0.461. Possible enhancement in this performance can be achieved by using more precise 3D models or maps. For instance, the 3D city models used here were of a medium accuracy of about 2–3 m in height estimation. More accurate 3D models at 1 m or better are available but at an extra cost. Better 3D maps can also be established from laser scanning, particularly in urban areas.

5.3 Accuracy and integrity monitoring results

Positioning in the above test was carried out using a system of combined RTK and IMU + odometer, where the latter method was required during gaps in RTK positioning spanning only short periods of up to 4 s, totaling about 3% of the entire test period. The position errors, computed as the difference between the solution from each of the positioning methods and the solution from a more precise system as explained in the previous section, are illustrated in **Figure 7**. **Table 3** shows the median of the absolute positioning errors and the RMSE for each mode. The median is used as it is less affected by outliers and skewed values of *PLs*. The table and **Figure 5** show that the RTK with correct ambiguity resolution provided positioning errors of a few cms and IMU + odometer provided sub-m level accuracy reaching 0.53 m after 4 s, which can grow to more than 2 m in less than 20s if left without

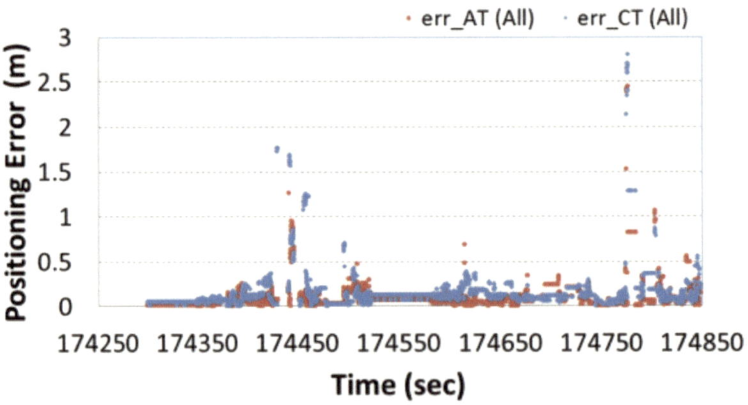

Figure 7.
Difference between using observed and predicted satellites in terms of: (left) number of satellites and (right) geometry expressed by PDOP.

Positioning mode	Median Error		RMSE	
	(AT)	(CT)	(AT)	(CT)
RTK	0.057	0.055	0.078	0.106
IMU + odometer within a few seconds	0.151	0.320	0.248	0.205

Table 3.
Median positioning error and RMSE for AT *and* CT *directions (m).*

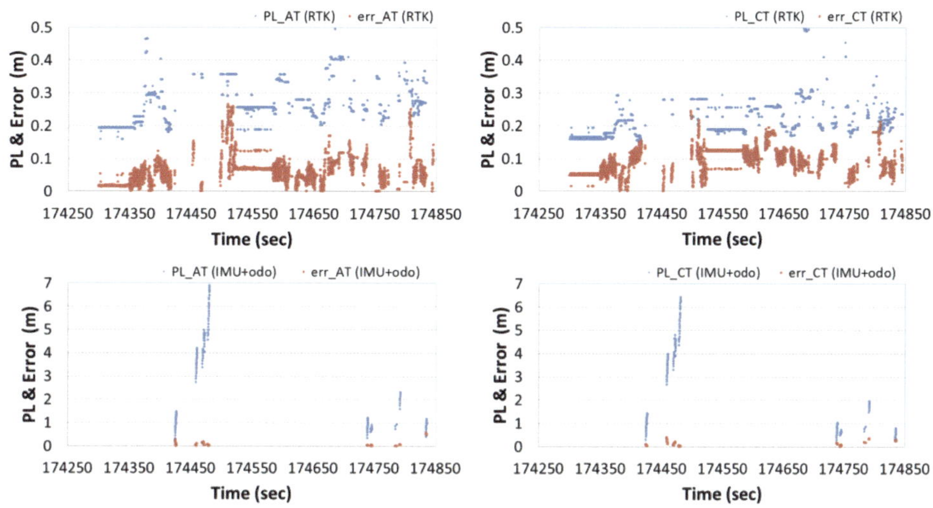

Figure 8.
PL_AT, PL_CT and PE for the AT (left) and CT (right) for the RTK (top panel), and IMU + odometer positioning (bottom panel), integrity risk = 1×10^{-5}.

calibration. Hence, this method should be limited to bridging RTK positioning only for a few-second period.

Figure 8 shows the time series of the *PL* when the RTK (top panel) and IMU + odometer (bottom panel) were used for the AT and CT directions (shown as *PL_AT* and *PL_CT*) using an integrity risk of 1×10^{-5}. The absolute values of the positioning errors in the *AT* and *CT* directions, i.e. *err_AT* and *err_CT* are depicted in the figure. Note the different scale used. The effectiveness of integrity monitoring algorithm can be assessed by checking that the position errors (*PEs*) are bounded by *PLs*, and availability of integrity monitoring is judged by checking that $PL < AL$. The RTK positioning errors shown in the figure were less than 10 cm and were always bounded by tight protection levels and thus the choice of an alert limit (*AL*) of 1.5 m (e.g. half of a small lane width of 3 m) is sufficient. Likewise, for the majority of IMU + odometer positioning period, the position errors were bounded by the PLs. Thus, positioning integrity was available (i.e. $PL < AL$) for the full period of RTK positioning and during most of IMU + odometer positioning, giving a total availability of integrity monitoring >99.9%.

6. Testing SBAS for ITS

To evaluate the next generation SBAS for ITS applications, Kinematic tests were conducted using two test vehicles in scenarios characterized by different levels of sky-visibility that may be experienced by ITS applications, which included open

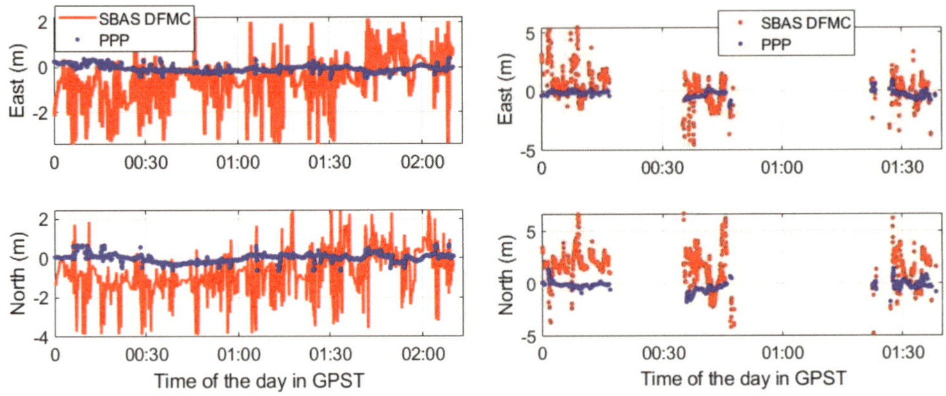

Figure 9.
Results of DFMC SBAS and SBAS-based PPP for open-sky environment (left) and semi-urban environment (right).

sky, semi-urban and urban environments. The tests were carried out in July and August 2018 in Australia. The next-generation SBAS test-bed data was used which provides three types of solutions, L1 SBAS solution, a DFMC SBAS solution using L1/L2 GPS and Galileo E1/E5a measurements, and SBAS-corrections for PPP solution using GPS and Galileo observations (L1/L2 and E1/E5a, respectively). All data were collected and processed at a frequency of 1 Hz. A 'ground truth' for the positions of the vehicle was computed in a post-processing mode through independent relative kinematic positioning - PPK. The same raw code and phase observations used for SBAS-based positioning were used as the rover observations with data from a Continuously Operating Reference Station (CORS), serving as a base station, where the test vehicles were within a radius of eight kilometers from the base station. Only ambiguity-fixed solutions from PPK, with 1–5 cm precision, were used as ground truth.

Figure 9 shows two examples of applying the new generation DFMC and float-ambiguity PPP SBAS in the open-sky (left panel) and semi-urban (right panel) environments. Analysis of results shows that the new SBAS DFMC solutions have slightly better accuracy than the SBAS L1 solutions, but both generate errors between sub-m to more than 2 m. In addition, the multi-constellation PPP solutions have shown to provide the best positioning precision and accuracy among all the tested solution types (L1 SBAS, DFMC SBAS, and SBAS-based PPP) with sub-decimeter level standard deviations after solution convergence [35], provided that enough convergence time is available, which may take up to 30 min. Testing clearly demonstrates that positioning performance of both DFMC SBAS and PPP methods is strongly dependent on the environment of the application, which is linked to the strength of the satellite geometry, number of observed satellites, the presence of multipath and NLOS. When considering the suitability of these findings for ITS applications, results suggest that, in open sky environment, lane identification and collision alert applications can be performed 80% of the time with DFMC SBAS and nearly all the time with PPP. The urban environment was the least promising with low availability for all SBAS solution types.

7. Cooperative positioning tests

A series of experiments with an aim to test the performance of different CP architectures were conducted in the vicinity of Melbourne, Australia. A network of

Figure 10.
CP equipment setup—Reference (left) on the car (right).

platforms was employed, with two vehicles acting as moving rovers and two static platforms acting as RSUs, shown in **Figure 10**. The tests were conducted in multiple environment scenarios such as in open sky, residential and dense urban environments. In all, five main sensors were deployed and attached to these platforms, including GNSS, IMU, VSS, UWB and DSRC. The deployment of these sensors on each platform is as described in **Table 4**.

A more suitable architecture for CP in a VANET is one that is decentralized in nature. The proposed architecture is depicted in **Figure 11** (top) where the different color lines represent different input/output similar to the centralized architecture in **Figure 10** (bottom). It can be seen that in this system, each vehicle only uses connections with immediate neighbors, i.e., data does not need to be hoped to a central processor in an event where direct connections between the central processor and vehicles cannot be established. This makes a VANET with a decentralized architecture more efficient in handling scalability compared to a centralized system. However, the nature of a decentralized system disallows for the computation of the network joint posterior, where only marginal posteriors can be calculated. Thus, in terms of the parameter estimation process, it is expected that the decentralized architecture and algorithm would produce less accurate navigation output compared to the centralized architecture. **Figure 11** shows that apart from the absence of a central processor, the decentralized architecture is identical to the centralized one. As before, the local level observations are solved for first, before the resulting information is passed on to the CP processor, utilizing SPAWN or MDSPAWN as

Sensor	Vehicle 1	Vehicle 2	RSU 1	RSU 2
GNSS	Leica GS10	Leica GS10	Leica GS15	Leica GS15
IMU	GX3-45	MMQG	-	-
VSS	Vehicle OBDII	-	-	-
DSRC	MK2	MK2	MK2	MK2
UWB	1 × P410	2 × P410	2 × P410	2 × P410

Table 4.
Different Sensor deployment for CP testing.

Figure 11.
Decentralized CP (top) and centralized CP (bottom), from [40].

the CP state estimator. In this example, the algorithms used for the centralized processing were the standard extended Kalman filter (EKF) [36] and particle filter (PF) [37] as well as an algorithm developed here termed the measurement directed progressive correction (MDPC), which is an improved centralized PF based on [37] for cooperative positioning in difficult environments. The decentralized approaches implemented were the sum product over wireless network (SPAWN) method, which is a fully distributed estimation algorithm, based on a factor graph (FG) and sum product algorithm (SPA) [38, 39]. This was further developed in this research and adopts the measurement directed method on SPAWN, termed the measurement directed SPAWN (MD-SPAWN) algorithm. Full details of these algorithms can be found in [40].

Results obtained from integrating INS, GNSS using real datasets where both non-radio ranging and radio ranging based CP were utilized are summarized in **Table 5**. The table shows that fusing INS and shared raw GNSS observations to form V2V double difference ranging in CP could significantly enhance the positioning accuracies of the participating vehicles in VANET. INS, which provided measurement redundancies and continuous positioning, enables for the CP to continuously operate even when the observed satellites were below the minimum number of

Scenario 2.1	Mean RMSE (m)		Max Error (m)	
	2D	3D	2D	3D
INS/GNSS	2.59	4.77	4.46	10.92
INS/GNSS CP EKF	1.21	2.22	2.45	4.24
INS/GNSS CP MDPC	1.69	3.19	3.20	5.63
INS/GNSS CP SPAWN	1.67	3.24	3.19	4.98
INS/GNSS CP MDSPAWN	1.63	3.09	3.08	4.71

Table 5.
Integrating INS with GNSS in different CP scenarios.

satellites needed in conventional CP system, which is useful in environments that suffer from GNSS signal shadowing. However, this technique is not suitable to be used when the GNSS signals are affected by multipath, which would invariably affect the V2V ranging. When radio ranging was obtained via UWB transceivers, instead of shared raw GNSS observations results obtained were comparable.

8. Conclusion

A proposed system that can be used for ITS is discussed. The system includes low-cost GNSS RTK integrated with MEMS IMU and automotive odometer. This low-cost RTK system can produce an accuracy of less than 0.1 m. The Integrity monitoring (IM) approach for this system is presented. The computed PLs were shown to bound the position errors all the time, proven effectiveness of the models. The PLs also suggested that the use of an alert limit of 1.5 m will provide IM availability larger than 99.9%. This can support several ITS applications. A main challenge in positioning using IMU + odometer is that they estimate the time-change in positioning and their biases accumulate with time, the system thus needs frequent resetting. Their bridging of positioning is thus recommended only for a few seconds, where sub-meter accuracy can be obtained.

The next-generation SBAS was also evaluated for ITS applications. It is evident from testing that the positioning performance of both DFMC SBAS and PPP methods is strongly dependent on the environment of application, where SBAS seems only suitable for open sky environment where DFMC SBAS and PPP can provide sub-m/m and dm accuracy, respectively.

To ensure safety, reduce pollution and save energy, reliable positioning with best route selection capability is required for intelligent transport systems. A method for route selection with the capability of predicting precision and integrity of positioning using GNSS integrated with 3D city models is presented. A demonstration of the method was conducted in a semi-urban area in Tokyo. Results show that the 3D city model based algorithm is able to determine LOS satellites with a small margin of error. An improvement can be achieved, for instance, by using higher accuracy 3D models or digital laser scanning 3D models. The method proposed is also suitable to determine locations within the city, where GNSS would not be suitable and other sensors should be used.

Future work includes the use of Lidar sensors and better quality IMUs, and developing new algorithms for the computation of the PLs for these sensors. Additionally, advanced models for weighting GNSS observations under multipath are considered. Finally, more architectures of V2V and VANET are planned.

Acknowledgements

FrontierSI, and Geoscience Australia (GA) are acknowledged for supporting SBAS data gathering through funding the SBAS project PD8703. Norman Cheung and Dr. Joon Wayn are acknowledged for their help in SBAS data collection. Dr. Azmir Hasnur Rabiain and the members of FIG working Group 5.4 and IAG sub commission 4.2 for their assistance in the data collection and data processing.

Author details

Ahmed El-Mowafy[1*], Nobuaki Kubo[2] and Allison Kealy[3]

1 School of Earth and Planetary Sciences, Curtin University, Australia

2 Tokyo University of Marine Science and Technology, Japan

3 School of Science, RMIT University, Australia

*Address all correspondence to: a.el-mowafy@curtin.edu.au

References

[1] Godha S, Canon ME. GPS/MEMS INS integrated system for navigation in urban areas. GPS Solutions. 2007;**11**(3): 193-203

[2] Yang Y, Mao X, Tain W. A novel method for low-cost IMU aiding GNSS attitude determination. Measurement Science and Technology. 2016;**27**(7): 075003

[3] Yand L, Wu Y, Li Y, Rizos C. An enhanced MEMS-INS/GNSS integrated system with fault detection and exclusion capability for land vehicle navigation in urban areas. GPS Solutions. 2014;**18**(4):593-603

[4] Lu N, Cheng N, Zhang N, Shen X, Mark JW. Connected vehicles: Solutions and challenges. IEEE Internet of Things Journal. 2014;**1**(4):289-299

[5] El-Mowafy A. Pilot evaluation of integrating GLONASS, Galileo and Bei Dou with GPS in ARAIM. Artificial Satellites. 2016;**51**(1):31-44

[6] Blanch J, Walter T, Enge P. Optimal positioning for advanced RAIM. Navigation. 2014;**60**(4):279-289

[7] Rippl M, Martini I, Belabbas B, Michael M. ARAIM operational performance tested. In: Flight Proceedings of ION ITM 2014, San Diego, CA; 27–29 January, 2014. pp. 601-615

[8] El-Mowafy A, Yang C. Limited sensitivity analysis of ARAIM availability for LPV-200 over Australia. Advances in Space Research. 2016;**57**(2): 659-670

[9] Khanafesh S, Pervan P. New approach for calculating position domain integrity risk for cycle resolution in carrier phase navigation systems. IEEE Transactions on Aerospace and Electronic Systems. 2010;**46**(1):296-306

[10] Cezón A, Cueto M, Fernández I. Analysis of multi-GNSS service performance assessment: ARAIM vs. IBPL performances. In: Proceedings of the ION GNSS 2013, Nashville; 16–20 September, 2013. pp. 2654-2663

[11] Wang K, El-Mowafy A. Integrity monitoring of intelligent transport system using Kalman filter based RTK. Journal of Geodesy. 2019. In Press

[12] Margaria D, Falletti E. A novel local integrity concept for GNSS receivers in urban vehicular contexts. In: Proceedings of the IEEE/ION PLANS 2014, Monterey, CA, USA; 5–8 May, 2014. pp. 413-425

[13] Santa J, Ubeda B, Toledo R, Skarmeta AFG. Monitoring the position integrity in road transport localization based services. In: IEEE Vehicular Technology Conference, Montreal, Quebec; 2006. pp. 1-5

[14] Zhu N, Betaille D, Marais J, Berbineau M. GNSS position integrity in urban environments: A review of literature. IEEE Transactions on Intelligent Transportation Systems. 2018;**19**(9):2762-2778

[15] Imparato D, El-Mowafy A, Rizos C, Wang J. A review of SBAS and RTK vulnerabilities in intelligent transport systems applications. In: Proceedings of the IGNSS Symposium 2018, Sydney; 7–9 February, 2018. pp. 1-18

[16] Hsu L, Gu Y, Kamijo. SNLOS correction/exclusion for GNSS measurement using RAIM and city building models. Sensors. 2015;**15**: 17329-17349

[17] El-Mowafy A, Kubo N. Integrity assessment of vehicle positioning for journey planning in urban environment using RTK and 3D city models. Journal of Navigation. 2019. In Press

[18] El-Mowafy A. Chapter 7: Precise real-time positioning using network RTK. In: Global Navigation Satellite Systems: Signal, Theory and Applications. Rijeka, Croatia: InTech Publisher; 2012. pp. 161-188

[19] El-Mowafy A, Wang K. Second generation SBAS performance analysis and bridging positioning and integrity monitoring during SBAS outages. In: Proceedings of the ION GNSS+ 2019, Miami, FL; 16–20 September, 2019. pp. 2842-2854

[20] Misra P, Enge P. Global Position System: Signals, Measurements, and Performance. Lincoln, Massachusetts: Ganga-Jamuna Press; 2006

[21] Zumberge JF, Heflin MB, Jefferson DC, Watkins MM, Webb FH. Precise point positioning for the efficient and robust analysis of GPS data from large networks. Journal of Geophysical Research. 1997;**102**(B3): 5005-5017

[22] Kouba J. A Guide to Using International GNSS Service (IGS) Products. 2015. Available from: https://kb.igs.org/hc/en-us/articles/201271873-A-Guide-to-Using-the-IGS-Products [Accessed: June 2019]

[23] Mahmoud A, Noureldin A, Hassanein H. Integrated positioning for connected vehicle. IEEE Transactions on Intelligent Transportation Systems. 2019. DOI: 10.1109/TITS.2019.2894522

[24] El-Sheimy N, Hou H, Niu X. Analysis and modeling of inertial sensors using AV. IEEE Transactions on Instrumentation and Measurement. 2008;**57**(1):140-149

[25] El-Mowafy A, Kubo N. Integrity monitoring of vehicle positioning in urban environment using RTK-GNSS, IMU and speedometer. Measurement Science and Technology. 2017;**28**(5): 055102

[26] Austroads. Cooperative ITS Strategic Plan. Austroads Research Report AP-R413-12. Austroads: Sydney; 2012

[27] Noah J, Smith B, Park B. Traffic signal control with connected vehicles. Transportation Research Record: Journal of the Transportation Research Board. 2013;**2381**(1):65-72

[28] Richter E, Obst M, Schubert R, Wanielik G. Cooperative relative localization using vehicle-to-vehicle communications. In: Proceedings of the 12th International Conference on Information Fusion, Seattle, WA, USA; July 2009. pp. 126-131

[29] Möller T, Trumbore B. Fast, minimum storage ray-triangle intersection. Journal of Graphics Tools. 1997;**2**(1):21-28

[30] El-Mowafy A, Kubo N. Integrity monitoring for positioning of intelligent transport systems using integrated RTK-GNSS, IMU and vehicle odometer. IET Intelligent Transport Systems. 2018; **12**(8):901-908

[31] Rife J, Pullen S, Enge P, Pervan B. Paired overbounding for nonideal LAAS and WAAS error distributions. IEEE Transactions on Aerospace and Electronic Systems. 2006;**42**(4): 1386-1395

[32] Chalko TJ. Estimating accuracy of GPS Doppler speed measurement using speed dilution of precision parameter. NU Journal of Discovery. 2009;**6**:4-9

[33] Zhao Y. GPS/IMU integrated system for land vehicle navigation based on MEMS [Thesis]. Stockholm, Sweden: KTH Royal Institute of Technology; 2011

[34] Meguro J, Kojima Y, Suzuki N, Teramoto E. Positioning technique based on vehicle trajectory using GPS

raw data and low-cost IMU. International Journal of Automotive Engineering. 2012;**3**(2):75-80

[35] El-Mowafy A, Cheung N, Rubinov E. Performance analysis of using the next generation Australian SBAS with precise point positioning capability for ITS. In: Proceedings of the ENC2019, Warsaw, Poland; 9–12 April, 2019. pp. 1-8

[36] Welch G, Bishop G. An Introduction to Kalman Filter. Chapel Hill: Department of Computer Science, University of North Carolina; 2006

[37] Morelande M, Moran B, Brazil M. Bayesian node localisation in wireless sensor networks. In: IEEE International Conference on A, S and S Processing, 2008 (ICASSP 2008); March 2008. pp. 2545-2548

[38] Caceres M, Penna F, Wymeersch H, Garello R. Hybrid cooperative positioning based on distributed belief propagation. Selected areas in communications. IEEE Journal. 2011; **29**(10):1948-1958

[39] Wymeersch H, Lien J, Win M. Cooperative localization in wireless networks. Proceedings of the IEEE. 2009;**97**(2):427-450

[40] Rabiain AH. Robust cooperative positioning for VANETs based on multi sensor system & enhanced estimation algorithms [PhD dissertation]. Melbourne, Australia: University of Melbourne; 2018

Chapter 4

Electrification of High-Mileage Mobility Services in Cities and at Airports

Josh Sperling and Alejandro Henao

Abstract

High-mileage vehicles serving airports offer significant potential for the electrification of transportation, in ways that enhance the affordability and sustainability of mobility for people and electric vehicle infrastructure development. As one example, by mid-2018, transportation network company (TNC) electric vehicles (EVs) in California—as a high-mileage mobility-as-a-service (MaaS) vehicle platform—was estimated to represent 30% of total non-Tesla EV charging demand, despite being only 0.5% of EVs in the State, and having sixty times higher levels of charging energy demand relative to the other EVs. This chapter explores the potential importance of this phenomena, the emergence of urban electric mobility developments and the co-benefits for economic, environment and equity. Through focus on the synergies of electrification with shared-use vehicles and trips, and with mobility options that include higher mileage, utilization, and occupancy, this chapter identifies emerging concepts that will have potential for impacting adoption rates, management, modeling and control for urban electric mobility systems. More specifically, this chapter explores emerging trends at and adaptations for airports. City airports, as critical hubs for TNC trip demand, and engines of regional economic growth, may be a critical locale for siting fast-charging infrastructure and planning new urban electric mobility operations across many metropolitan areas and cities of the United States and globally.

Keywords: new mobility choices, efficient and electric mobility infrastructure systems, airports and urban infrastructure modernization, shared-smart e-mobility, electrification and urbanization dynamics

1. Introduction: a convergence of megatrends for U.S. airports and roles in advancing smart, resilient urban electrification

The focus of this chapter is on the convergence of advanced electric mobility technologies and mobility services that are in use today. This focus has the potential to inform futures of connected and automated vehicles, to fueling/charging of electric mobility systems, to planning and designing for new choices, passenger travel behaviors, and infrastructure-related design and operations decisions. While research has lagged behind the rapid pace of disruption in the mobility marketplace, research has clearly identified urban contexts as where most disruption has happened so far– and where experimentations with new forms of integration of transportation, electricity, and other built systems have the potential to enhance access to economic EV pathways and other sustainability opportunities (across cities and rural communities alike). As an important hub for travel, airports are identified as often being the "front doors" for accessing most cities, and with these hubs helping connect the city to the world. The impacts of these systems are important for understanding the opportunities associated with electrification of high-mileage, shared mobility services, that are increasingly available at airports or within/across cities.

We start with a recent finding: that by mid-2018, high-mileage, transportation network company (TNC) electric vehicles in California were estimated to account for over 35% of total non-Tesla EV charging demand, despite being only 0.5% of EVs in the State [1]. While the use of TNCs at pickup and dropoff points within cities and airports have become increasingly observable, and creating new challenges for curbside management—this initial finding on the key EV charging aspects (that will relate to future curb uses) offers a useful overall chapter rationale and motivation. Similar to TNC fees and revenues, to parking and car rental revenues being of primary interest to airports, the electrification of high-mileage urban mobility services such as TNCs are of prime interest to electric power utilities and cities. Its within this context that we explore related and interdependent questions for informing future e-mobility and energy efficient management of mobility, with new data-driven discovery, behavioral models, and control techniques (e.g., of curb management, fee structures, vehicle caps, and other incentives) that are emerging in cities and especially at airports with TNCs. Aligning with the scope of the book, this chapter focuses on synergies of EV shared mobility inclusive of hybrid and EV technology for intelligent, efficient transport, especially within contexts of developing energy efficient mobility management, modeling, and control techniques. The need for examining energy-efficient, smart cities, and mobility systems innovations—defined by cities that harness and validate using data, new technology, and governance strategies—and related future grid/infrastructure impacts, is based a priority of improving the efficient, affordable, and reliable movement of people, while reducing costs, environmental impacts, congestion, and enhancing infrastructure modernization.

The three key questions posed in this chapter to inform future modeling is around urban electrification priorities at airports and emerging sustainability goals, and include the following:

1. How does research and data-driven insights on EV-shared mobility or new MaaS/TNC choices (by consumers, systems designers/operators, to diverse system management actors) support understanding of urban electrification, electric vehicle (EV) grid impacts, EV-shared mobility infrastructure, behavior change, and technology adoption for future mobility systems?

2. How do new mobility choices, energy-efficient or intelligent transportation system management strategies, infrastructure modernization, and new business models help inform the research, development, demonstration (R2D) and commercialization of (hybrid) electrified urban transport and mobility systems, and within the context of utility industry-led business models?

3. How does R2D approaches on new mobility choices, infrastructure, to new revenue, finance and business models support development of transport system modeling, management and control—inclusive of new disruptions of transportation network company vehicle electrification?

Trends from January 2016 to 2018 show that the company Uber has seen growth from 1 billion to over 10 billion trips, with Lyft seeing an uptick from 100 million to 1 billion rides. In fact, latest figures have revealed that almost 15% of all rides are trips to and from airports and almost 10% of total city vehicle miles traveled (VMT) can be to and from an airport. This chapter explores the rapid growth and initial impact dynamics of Uber, Lyft, and other forms of transportation network companies (TNCs) at airports and in cities, and related megatrends that might shape the electrification of high-mileage mobility services in cities and at airports [1]. The four primary megatrends described in this chapter, include: (1) new mobility as a service (MaaS) transitions (led by the proliferation of on-demand ride-hailing services and TNCs), (2) the integration of new mobility and energy (or electrification) choices, (3) investments in increasingly electrified, intelligent, and energy-efficient mobility infrastructure modernization, and (4) revenue diversification, enabled by adaptations by airports to TNCs, and that may also help to inform future mobility electrification strategies by utilities, drivers, and cities. As new mode choices and comparative insights emerge, this chapter builds on initial quantification of shifts in ground transportation at airports, to identify ways in which adaptations to new mobility as a service (MaaS) options may offer opportunities to significantly disrupt decade old patterns of private gas vehicle ownership, as well as accessibility to or affordability of EV-shared mobility. By exploring key destinations and hubs for these new mobility options (e.g., airports, cities), this chapter aims to learn from emerging trends by integrating data, literature and analytical insights. The trends identified motivate further exploring and modeling of emerging impacts at airports and cities, and so to inform the enabling of hybrids (HEVs), plug-in hybrid (PHEV), and battery electric vehicles (BEVs) and varying urban electrification future at multiple rapidly growing airports. The future research directions and questions identified in this chapter offer initial lessons and knowledge to shape the future of harnessing urban systems data integration for modeling, energy-efficient management and new control algorithms for ride-hailing fleets of automated vehicles that are both electric and shared (by multiple travelers).

1.1 Chapter rationale

The US and global urban transport and mobility system is experiencing significant disruptions, starting over the last two decades in areas of car/bike sharing (e.g., ZipCar, Capital bike share, Divvy, CitiBike, Bixi, Mobike; [2]) and now increasingly in peer-to-peer vehicle and ride-sharing (e.g., Waze Carpool, GetAround; [3]), on-demand ride-hailing services as transportation network companies (TNCs—e.g., Uber, Lyft, Via, Didi, Grab, Ola, 99; [4]), micro-transit, micro-mobility as dockless e-scooters and bikes (e.g., Lime, Jump, Ofo), and as shared EV mobility services. For EV-shared mobility (see EVsharedmobility.org), the trends emerging are clear: new uses of smart-phones and app-based approaches to ride-hailing is enabling entirely new mobility choices (especially observable at airport), EV charging infrastructure is enabling EV adoption (e.g., EVGo, a smartphone app for locating more than 1200 fast-chargers in 66 metropolitan markets and is now charging its network with 100% renewable energy); to TNC drivers using EVs or plug-in and hybrid EVs for saving on fuel costs and maintenance over time. In fact, investments by many automotives such as BMW-Daimler's ShareNow, Ford's Office Ride or GM's MavenGig, have been some of the first to offer EVs for rent to TNC ride-hailing drivers. This is coupled with EV battery prices of more than $1200 per kWh in 2005 dropping to be as low as $200 (or even $125) per kWh for some suppliers in 2020–2025 [5]. All these trends are leading to whole new models and possibilities for daily driver and passenger choices around both mobility and energy.

At the same time, air transportation is among the fastest growing modes (in terms of passenger miles traveled) within the transportation sector, with TNC demand rising at airports, and global air traffic over past 20 years increasing at a rate of 5–6% annually (indicating a doubling time of every 12–14 years) [6]. In the U.S., airports have been vital national resources—enabling movement of people and goods to regional, national and international commerce—yet have also faced significant and growing congestion problems, particularly with increased curb demand and competing uses for right of way within major areas with continued population, economic growth [7], and several new mobility choices.

As regional examples, aviation demand in the southern California region, including San Diego, Tijuana, and five airports in the Los Angeles area, is projected to increase 50% between 2009 and 2030, from 48 to 80 million passenger enplanements [8]. In the New York City metro region, served by JFK, La Guardia and Newark airports, limited efforts have been made to create multi-airport regional demand modeling forecasts and perhaps due to growing concerns on the ability of traditional demand modeling techniques (including multinomial logit formulations for examining airport choice) to produce reliable air traffic and airport-level demand forecasts (especially with uncertainty due to multiple factors, e.g., fuel prices) [9]. Meanwhile, and as of December 2016, new transportation network companies (TNCs—e.g., Uber, Lyft) have been permitted to operate at more than 90 U.S. airports, with some airport data on effects and new strategies—e.g., TNC service revenues (87% of surveyed airports requiring a per-trip service fee), public parking and rental car revenues [10], to terminal building roadway operation design focused on allocating curb space to use of high-occupancy or the most-efficient services [11], and managing traffic volume increases associated with new TNC travel mode shifts [12]. More recently, LAX airport has decided to end their airport curbside pickup of travelers for Uber and Lyft, and instead have passengers travel by shuttle to a nearby parking lot with less congestion, noting "the decision is in response to worsening congestion at the airport, which is undergoing a $14 billion overhaul of its aging road network and terminals" [13]. Similar changes have occurred for San Francisco Airport and will soon take place for Boston's airport as well; opening up new economic questions of airport curbs as low supply, with high demand (enabling additional new revenue opportunities), with research now exploring intelligent, efficient control options such as premium, regular, and budget access for ride-hailing with different fees based on more or less convenient access (including perhaps for vehicles that are higher occupancy, so less congestion; and with higher fuel economy or clean energy electric vehicles, so less air pollution).

Furthermore, airports are increasingly becoming the primary driver determining mobility and economic development trends, particularly in vibrant urban areas. Just as coastlines, river travel, railroads, and highways greatly influenced the design and landscape topology of cities and urban mobility networks, air travel is now exerting its influence in urban development. Mid-size cities such as Denver see a large percentage of development focused on air-front access, while large mega cities (Chicago, New York, LA) are turning their attention on connecting their multiple airports as efficiently as possible to the urban core. Indeed, TNC/taxi drivers are also seeing highest revenues and passenger trip demand within cities being at their airports [14].

1.2 Four megatrends: rapid changes in mobility, energy, infrastructure and finance

Within this context, this paper explores four inter-related megatrends at airports as changing multiple industries (especially parking and curbs—as noted in the airport case study review section). **Figure 1** illustrates a summary of the key

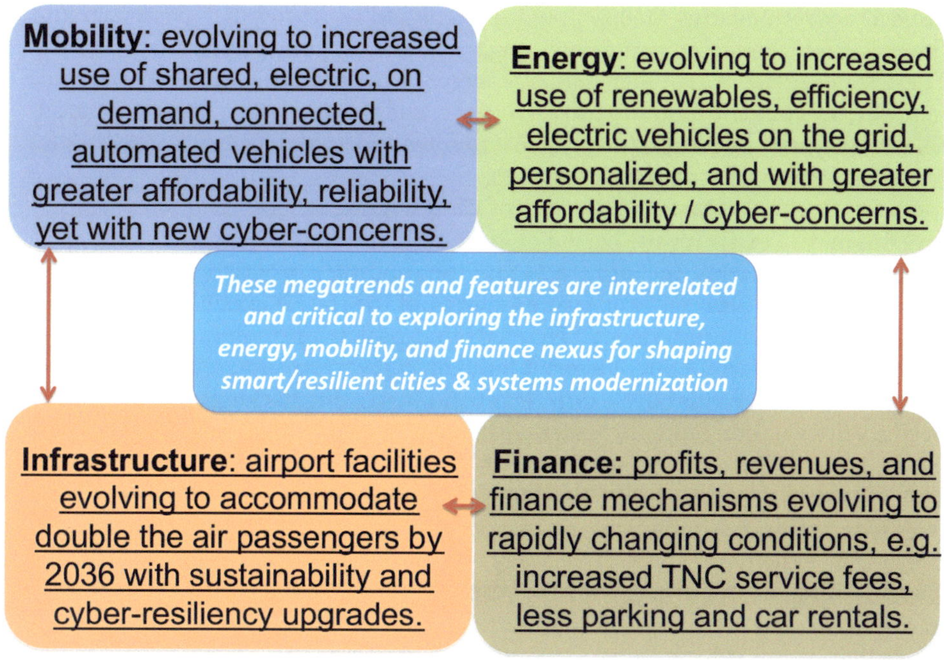

Figure 1.
Interdependencies of mobility, energy/e-mobility, infrastructure, and finance for airports.

interactions, interdependencies, and synergies of these important trends as new urban mobility, energy (including electric utility revenues associated with increasing electric mobility), infrastructure, and finance/revenue pathways unfold, with the learning from data-driven discovery of insights at airports as trends that could inevitably impact cities too. The contexts of global, national, and city levels—based on literature, data, and analytical insights—is shared next for enabling smart cities and urban systems modernization that focus on harnessing high-mileage, ride-hailing MaaS or TNC electric vehicles and shared mobility for enabling accelerated urban electrification, defined in this chapter as:

> *the process of powering [transportation systems] by electricity and, particularly in cities in advanced economies, the introduction of such power by changing over from an earlier power source (see [15]).*

1.2.1 Key message up front: a smart mobility "leapfrog" is a move beyond privately owned EVs to shared e-mobility in urban and airport areas

The term "leapfrogging" has been applied to cities and nations that have adopted a new form of infrastructure by bypassing the traditional progression of development, e.g., from no phones to cell phones—bypassing landlines all together. For the first time in history, similar transformations (perhaps even "leapfrogging" opportunities) may exist for shared, energy-efficient, and e-mobility services in cities that can reduce congestion and pollution while increasing service affordability and "private car-optional" multi-modal accessibility.

According to recent analyses, travel facilities are expected to host 7.8 billion air passengers by 2036, up from 4 billion air travelers today [16]. According to Airports Council International-North America, Airports in the United States are anticipated to have a collective funding need of more than $75 billion over the period of 2015–2019, or $15 billion per year to successfully complete infrastructure projects accommodating facility upgrade needs and high growth in passenger and freight activity-inducing many new congestion and air pollution issues [17, 18]. These funding needs are much higher if there is a need to build out more parking, roadway, and electric mobility infrastructure, and this is where a focus on shared EV mobility may be critical towards cost savings and as high priority measures for reducing pollution for airports and cities.

According to the American Society of Civil Engineers, a two trillion-dollar investment gap exists for the period of 2015–2025 for maintaining our existing infrastructure systems, including a $42 billion funding gap between 2016 and 2025 for aviation. This also comes at a time when 24 of the top 30 major U.S. airports are expected to soon experience "Thanksgiving-peak traffic volume" at least 1 day every week. Similarly, 1 out of every 5 miles of highway pavement is already in poor condition and our roads have a significant and increasing backlog of rehabilitation needs. With aviation industry leaders seeing record $29.3 billion net profits (in 2015), up from $16.4 billion in 2014 (with many North American carriers responsible for over half these profits), leadership is also viewing new infrastructure upgrades as a key need to accommodate rising demands. US and global airports are continuing to focus on harnessing emerging technology services, infrastructure (re)development and investments that can advance local smart cities efforts too, in response to rapid urbanization (growth in urban populations) and motorization (growth in motor vehicle ownership—especially in Asia and Africa).

Integrating EV with shared mobility, as well as with high capacity and high-frequency transit, will enable the benefits of both less air pollution and congestion, creating win-win airport and city opportunities—as well as benefits towards affordability for travelers. With new mobility choices that are increasingly cost and time-efficient, as well as more comfortable services (perhaps relative to transit), energy-efficient or intelligent transportation system management strategies, infrastructure modernization, and new business models are needed that enable acceleration of higher occupancy and multi-passenger forms of (hybrid) electrified urban transport and mobility systems. Leading utilities, such as the New York Power Authority (NYPA), have identified airports (both JFK, and La Guardia; [19]) as primary locations for investment in high-speed electric vehicle charging infrastructure, specifically for Uber and Lyft (TNC) vehicles waiting to pick up passengers. LAX airport in Los Angeles has taken on similar procurement with high mileage bus fleets for airside operations, while recently removing access to the overcrowded curbside to reduce congestion around the terminals [20]. Such approaches have the potential to demonstrate how airports, as major traffic corridors with large fleets can play a leading role in enabling less congestion, and perhaps further incentivizing electrified shared mobility and public transit (allowing those modes easier, lower-cost access to the 'front-door or front curbside') rather than privately-owned gas vehicles or even private electric vehicles. Such considerations might send the right signals for transformation and catalyze new market investments and industry-led business models, initially through 'fast-charging' infrastructure deployments and appropriate utility rate structures to enable accelerated deployment and demonstration of high-mileage, and high-utilization vehicles with easy access to fast-charging infrastructure at airports. Using this model, cities may learn lessons as to best areas for siting charging infrastructure to accommodate similar forms of new on-demand EV mobility services.

While profits are up for airports—with the Federal Aviation Administration in 2016 estimating that $3.5 billion were collected in parking and ground transportation fees, representing 41% of the $8.5 billion in U.S. airport revenue not related to airlines—changes away from parking and car rentals at airports, partially due to TNCs, are motivating new revenue collection approaches as curb pickup-dropoff fees, and perhaps even EV fast-charging infrastructure for TNC drivers.

Some of these initial concepts are further outlined later in this chapter, as new business and finance models for airports, cities, electric utilities and states (ACES) to collect new revenues, reduce congestion, pollution and sustainable infrastructure investment opportunities with high-utilization EV charging infrastructure and electrification/incentivized charging of high-mileage TNC vehicles in cities. This may also help to inform transitions and transformations towards increasingly automated, connected, efficient/electric and shared mobility systems (ACES2).

1.2.2 City-level context

Growing cities and airports are facing complex challenges, ranging from a need to respond to growth in air travel, varied stresses and shocks (e.g., cybersecurity, natural hazards, aging infrastructure) to new disruptive advances in mobility, energy, and revenue mechanisms [21]. Technological disruption is especially true with respect to airport access and egress, where TNCs are making inroads in the percentage of airport passengers served, creating new curb demands and impacting traditional parking and car rental demands. Whereas tracing mode choices for the daily routine of commute to work is often governed by factors that need to be accounted for in a holistic approach to lifestyle and affordability, the access and egress trip to the airport and its associated mode can typically be determined on a per trip basis, taking into account factors such as the economic, efficiency, reliability, and convenience for only that trip (rather than a decision to invest multiple years in a vehicle mortgage). As such, airports offer initial front lines of observing new mobility trends, changes in access and egress patterns, both of which in turn are impacting revenue streams. If these indications foreshadow broader urban mobility trends, airports can be seen as the 'canary in the coal mine' with respect to larger urban scale impacts. A new critical question is with respect to the TNC rate of impacts for the airport versus the city. Currently, and as shown in the results of this chapter, available data still only exists for airports due to charging fees and public records requests. In the near future, anticipated pickup/drop-off (PUDO) or occupancy fee structures implemented across a few cities (e.g., as indicated by regulations emerging in New York, Chicago, Sao Paolo and other cities) may offer new insight.

At the same time, and with the rapidly growing demand for travel in cities, smart city leaders can learn from airports that are already facing significant difficulties keeping pace with technology and market disruptions and responding to unique new pressures and risks to their revenue growth, infrastructure investments and service user satisfaction. While airports in major U.S. cities already struggle to supply reliable access to terminals and manage impacts from private mobility service providers, increases in the types of stresses (e.g., rapid growth in demand for new private mobility services) and shocks (e.g., weather-related airport and power outages in Houston and Atlanta; to cybersecurity attacks on Atlanta's city departments) are also posing new challenges—that may be exacerbated by the surface of potential future cyber-attacks that may become available via an increasingly connected transportation and electric power grid system.

Cities often view their airports as critical and interdependent infrastructure systems for the regional economy, and therefore resilient airport operations—including maintaining quality of and reliability of services is a paramount concern—whether that's Atlanta, Georgia during and after airport power outages; Houston, Texas in recovering from flooding; Los Angeles, California in preparation for a large population influx for the upcoming 2028 Olympic Games.

These challenges identified, may also offer opportunities for airports to perhaps succeed where cities and others have not been proactive, especially in developing longer-term strategies. These responses can therefore offer lessons for city management in how best to respond to industry changes (e.g., decreasing parking revenues cities are also dependent on) to new approaches to finance modernization of aging infrastructure assets or upgrading of urban systems and facilities.

As to new decision environments or responses, the literature review presents a summary of megatrend evolutions and responses at 10 U.S. city airports, based on recent events and related technology services emerging. The diverse range of investment actions identified is discussed as having implications for smart and resilient urban environments, and city-level responses. Key factors and drivers are noted from data and literature-review of these city airports, their metro areas, and noting these EV-related disruption concerns or challenges. The chapter concludes with exploring the need for new predictive capabilities for forecasting EV demand, measuring effectiveness of strategies, and more integrated approaches across the four megatrends of: (1) business model disruption as new affordable, shared, electric, to automated mobility; (2) new cost-effective, distributed, and resilient energy for EVs; (3) airport-related economic opportunity driving infrastructure modernization; (4) revenue diversification and new finance demands.

2. Methods

This paper takes a two-pronged approach. First, a review of literature is conducted to inform analyses and structure the paper in terms of relevant EV mobility, energy or grid impacts, infrastructure and finance priorities. Second, initial data on TNCs at airports is discussed for exploring potential parallel and synergistic efforts for shaping futures of urban electrification.

2.1 An integrated systems review: definitions and assessments relevant to megatrends

Table 1 provides definitions *of* and key drivers of change *for* systems trends identified within the emerging literature and with airports and cities increasingly recognizing the need for building more integrated and resilient systems, that have the *ability to recover from or adjust easily to change*. These changes include adapting to and siting electric mobility infrastructure to support services.

Systems definitions	Airport systems	Urban systems
Transport/E-mobility: ***The quality or state of e-mobility system supporting efficient movement of people and goods and access; the ability/capacity to move in a system***	The Federal Aviation Administration (FAA) launched NextGen, **to modernize** the nation's air transportation and mobility system to increase safety, efficiency, capacity, predictability and resiliency of American aviation [22]. \$478 million in airport infrastructure grants to 232 airports in 43 states, as the fourth allotment of the total \$3.18 billion in FAA Airport Improvement Program (AIP) funding for airports across the United States is now underway.	The Department of Transportation focus on smart cities is in context of **how best to: move people and goods by supporting affordable, sustainable, safe, efficient mobility; adapt/be resilient to change** (e.g., TNCs, automation); **grow opportunity for all; and align decisions/ investments** [23]. To enhance affordability of EV/EVSE adoption, introduction to EVs as part of vehicle/ride sharing may enable greater population accessibility and affordability, relative to private vehicle ownership [24]. ~Half the costs for fueling an EV compared to conventional internal combustion vehicle

Systems definitions	Airport systems	Urban systems
Energy systems: ***usable power from resources transferred in production/ distribution systems for doing work***	A rise in energy to transport electrification needs and needs **to satisfy energy demands from: moving passengers and cargo; building, terminal and parking facilities; ensuring energy resilience perspectives; cost savings opportunities** via new forms of energy generation [25]	Urbanization and electrification has clear links to energy consumption, emissions and **key challenges include reducing peak demand power interruptions, ensuring energy security and resilience, and meeting rising demand for affordable, reliable, clean energy** [26]
Infrastructure systems (IS): ***underlying foundation, as of a system of public works of a country or region***	Over next 5 years, US airports may undergo $100s of billions in **infrastructure projects to accommodate growth in passenger and cargo activity, rehabilitate existing facilities and to support innovation**; airport access capacity in many areas may limit further airport growth without new solutions.	America's urban areas are often congested and traffic delays cost the country $billions in wasted time and fuel; **in urban settings, service disruption in one infrastructure system will almost always disrupt one or more other infrastructure systems** [27]. New infrastructure investments for electric mobility are now a priority.
Finance systems: ***the management of funds and systems for making/ aggregating investments***	"Although nearly all U.S. airports are owned by state of local governments, **airports are required to be as self-sustaining as possible, and receive little to no taxpayer support**" [28]; much of airport revenue is predicated on existing access/egress modes—namely all requiring parking.	"Urbanization is one of the most important potential drivers of productivity and growth in the global economy...**what's invested in today could lock in economic benefits or costs for decades to come**" [29]. Competitive advantage in attracting investment among cities is enabled by ambitious EV targets and goals.

Table 1.
Systems definitions/assessment of megatrends at airports and broader urban systems.

2.2 Airport and city reviews: disruptive shifts informing smart airport, city and urban systems modernization

By bringing together an understanding of intelligent mobility within context of shared use vehicles and rides, together with diffusion of EVs, cities and airports globally are beginning to respond and adapt to new shifts and opportunities for modernization. Examples include:

- Sao Paolo, where the city's new regulatory framework taxes ride hailing services at a lower rate for shared rides and electric cars, and for trips to parts of the city that have limited transport options;

- NYC is also considering a similar approach, $2.50 per trip for taxis, $2.75 for for-hire, $0.75 shared for-hire (note: this was recently updated to $2.50 for private ride-hail trips, and shared rides at $1.25 that will tentatively start in October, 2020), yet this is just for shared/higher occupancy (not incentivizing EVs), and with some pushback: e.g., "While the Albany plan will charge pooled passengers a smaller surcharge ($0.75), it doesn't go far enough to recognize the enormous benefits of ride-sharing and microtransit. In fact, on a percentage basis, riders in shared-trip vehicles will be paying a higher tax than those who choose to ride alone, and in some cases the total tax per vehicle will be higher too. During morning and evening commutes, Via transports six people per vehicle; the collective surcharge on those vehicles will be $4.50, larger than for lone riders in other for-hire vehicles and infinitely more than for the person driving his or her own car without charge." This is now being extended to the three major airports, Newark, LaGuardia and John F. Kennedy Airports for the Port Authority of New York and New Jersey, where approval has been made of La Guardia also offering a $15 flat fee LGA Connect service with Via (as another emerging ride-hailing operator) from LGA to most New York City boroughs, and farther trips as $20.

- Seattle Airport: "To meet environmental goals, the Port of Seattle only allows Uber vehicles with a blended MPG rating of 45 or higher to wait for airport requests at the waiting area" (where many TNC drivers are using vehicles of this nature that are hybrid electric vehicles such as Priuses).

Many forms of city experimentation to behavior change case study examples are emerging in this area. This will include a need to explore some of the unanticipated effects too—e.g., consumers choosing shared rides when they anticipate no one will actually share (or 'match').

For purposes of this next section of the review, urban systems modernization is now defined in the context of mobility, energy, infrastructure, and financial/economic systems (that include above examples of fee structures). For financial systems and key business shifts, this review includes emphasis on disruptions to parking demand and opportunities for land use conversion of a 2017 US$6 billion annual parking and car rental revenue industry. These cases are used to explore the emerging or potential performance metrics that might help to define key challenges, opportunities, and trends for informing decisions. Finally, the review notes and identifies key contexts, focused on enabling more resilient systems.

1. San Francisco International Airport (SFO)

   - *Mobility*: From 2015 to 2016, airport ride-hail trips increased 75% while airport BART ridership and parking demand decreased, leading to significant reduction in transit and parking fare recovery [30]. Nationally, in 2017, 68% of business travelers chose Uber or Lyft, while only 25% chose rental cars [31]. Since 2009, the city installed over 200 EV charging stations including at municipal garages and SFO airport, with anticipation of EV and gas-powered car prices being comparable as early as 2025 (for light weight, high-mileage vehicles). In the city, trips made by TNCs doubled from 2016 to 2017; and by the end of 2016, TNC vehicles were making over 170,000 trips within San Francisco per day, estimated as ~15% of all intra San Francisco trips, about 20% of VMT, and with ~1% of California TNC vehicles being EVs.

   - *Energy*: SFO consumes 440 GWh of energy each year and is one of the Bay Area region's largest energy consumers; SFO has the goal of achieving a zero net energy campus by 2021 and has decreased energy use by 1.5 GWh from a 2013 baseline despite expanding infrastructure/increasing passenger traffic; a 5% energy use reduction is estimated to save the airport $2.1 million per year [32]; "with air travel expected to double over next 20 years, many airports foresee future power supply constraints as they build new terminals and increase electric load. Due to location, SFO has limited options to bring in additional electric feeders and is exploring how on-site generation may help meet expected need" [33].

- *Infrastructure*: SFO has 5171 acres of land; 14.5 million square feet of space; 135 buildings; and over 200 plug-in electric vehicle parking stall locations, with free charging. For citywide infrastructure, the city currently accommodates over 413,000 registered vehicles, and this includes only 128,551 that are parked in single family home units (that have the most flexibility and lowest investment costs for EV charging installations); and a significant proportion as multi-unit dwellings without easy access to at-home or on-street charging.

- *Finance*: curb-side pickups/drop-offs for Uber, Lyft and Wingz increased from $3.80 to $5 on July 1, 2018 [34]; in Oct 2017, SFO TNC service fee monthly revenue: ~$3 million ($2,941,538), parking revenue: ~$9 million ($9,093,988) [10].

- *Resilience*: as part of the airport development plan utility projects, SFO is adding shoreline and sea wall protection for flood control and sea level rise; as well as new fuel storage tanks for extended outages and as contingency for fuel supply interruptions [32].

2. Denver International Airport (DIA)

- *Mobility*: to keep up with growing traffic, DIA has made significant investments including a $78.9 M purchase, that added 26 more cars (retiring 16 from airport opening in 1995) and increased reliability to a 23-year old automated guideway transit system connecting the terminal to all three concourses [35]. To address emerging pollution concerns city-wide, and as related to electrifying VMT for ride-hailing, the company, Lyft, launched the Express Drive rental program for TNC drivers, deploying 200 long-range EVs in this initial fleet (all of which, as part of this new rideshare rental program, received up to $5000 as an EV tax credit) and with plans to reach a 100% EV network.

- *Energy*: Denver also has over 300 electric vehicle charging ports and has been at the forefront of developing on-airport solar PV facilities, with four projects over 55 acres totaling 10 MW, with design generating capacity at 16.1 M kWh—enough electricity to power roughly 2580 Denver residences (owned/operated by a third party; power purchase agreement; excess power is sold to Xcel Energy) [36]. Falling costs have also motivated a new 1.6 MW PV carport on DIA 800-space parking lot and as a multi-stakeholder microgrid financed between Panasonic, Xcel Energy and DIA [37]. Total 2015 baseline energy use (1,250,801 MMBtu): 59% electricity; 41% natural gas; 18% terminal building; 15% Concourse B; 9% Concourse A; 7% Concourse C; 28% Central Plant; Other: 23% (e.g., 4% Cargo buildings; 4% parking; 3% train; 2% infrastructure) [38].

- *Infrastructure*: DIA identified as a "city within a city"—thousands of employees, 50 million-plus passengers annually; and a $544M South Terminal Redevelopment has included a public plaza, hotel/conference center (opened in 2015), and public transit center (opened in 2016). The city housing infrastructure conditions consists of 44% of residents living in multi-family housing; and just recently, plug-in electric vehicle ready building codes are expected to help address this challenge of increasing the availability and enabling of charging in multi-family buildings.

- *Finance*: In 2016, public parking provided $160 M in revenues; ground transport: $11 M; employee parking: $6 M [39]; TNCs: $3.2 M [38]. For EV charging, the Charge Ahead Colorado program has provides grant funding that covers up to 80% (up to $6260) of the cost of a Level 2 multi-port charging station. At the same time, the capital costs for DCFC stations in the Denver urban areas could be approximately $170,000, and with new funding sources needed to offset the high current operating costs of DCFC stations. The most economical prospects, due to high utilization, could prove to be for TNCs and the fleet of 200 EVs in the Lyft Express Drive fleet, that could more frequently utilize DCFC at airport facilities.

- *Resilience*: this was a criterion for evaluation for the airport energy master plan, referring to an ability of DIA energy systems to adapt to changing conditions, withstand disruption or recover quickly. Power failure metrics have included increasing reliability and resilience in ways that bring benefits for cost effectiveness and carbon reduction (e.g., microgrid/storage opportunities-harnessing renewables, off-peak utility rates, and increasing outage response capability) [38].

3. Kansas City International Airport (KCI)

- *Mobility*: the first 6 months of 2018 saw a 1–5% increase in total passengers over the same month in previous year. There were 11.5 M total travelers in 2017 and already 5.7 M travelers as of June 2018. For freight mobility, growth is even faster with 198,491, 890 lbs. of air freight moved through KCI in 2017 [40]. If to combine freight and mail transported by air, KCI experienced a 11.4% increase for year end 2016 versus 2015 [41].

- *Energy*: KCI became a first U.S. airport to integrate fully electric shuttles for passenger service with four BYD 30-foot coaches and the new KCI terminal has a goal of 100% renewable energy. In addition, the city—with Kansas City Power and Light—has invested in over 1100 charging stations.

- *Infrastructure*: 551 acres of pavement (as equivalent: 2125 road miles); another 52 miles of actual roads; 189 miles of airfield lighting wire; 23,000 parking spaces [42]; new $100 million KCI terminal (opening 2022) to replace the three existing terminals [43].

- *Financial/economic impact*: airport activity estimated in FY2015 contributed to 41,625 jobs; $1.41 billion in earnings; $5.02 billion in economic output to the 17-county primary service area; representing $2.79 billion value add to area's GDP (2.4% of area GDP) [44].

- *Resilience*: few details for airport; yet their DOT smart city vision Element #11: low-cost, efficient, secure, and resilient information and communications technology (ICT) [45].

4. Portland

- *Mobility*: in 1998, public transit made up 10% of access to the airport [11]; more recently, in 2017, and 2–3 years after transportation network companies have been serving PDX, TNC services make up over an estimated 10% of access to the airport for passengers [10]. As shown in **Figure 2** from 2017 City of

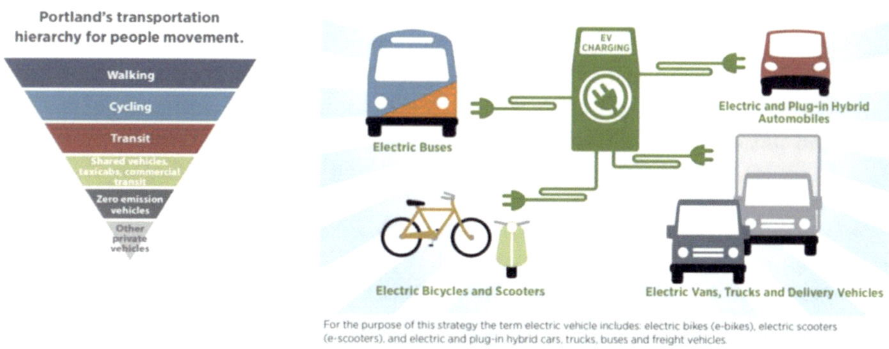

Figure 2.
Portland's prioritizing of people first for EVs (source: City of Portland [46]).

Portland Electric Vehicle strategy, the city aims to prioritize people movement over car movement, and so that includes focusing on electrification of the public transit system as a first priority, then shared vehicles and then private automobiles that remain in use—while also seeking to maximize benefits to air quality and affordability for low-income residents and parts of the city that are most dependent on private vehicles. Similarly, Portland became one of the first cities for Lyft to offer the option of 'Green Mode', with consumers being given the option to request EV rides. This is a significant step, yet with careful planning in entering additional city markets due to the limited supply of EVs in TNC fleets, and therefore the potential for longer consumer wait-time.

- *Energy*: Currently, and due to the low cost of electricity, an EV in Portland can travel a mile with electricity for about one third the price of gasoline [47]. In 2015, PDX had the largest installation of commercial electric vehicle chargers of any U.S. airport with 42 EV charging stations and is continuously expanding related infrastructure based on demand. By 2035, PDX aims to obtain 100% of operating power for PDX-controlled facilities from renewables and have in-building energy efficiency levels of 45 W/M2 [48]. By 2020, the City also has goals to increase access to EV charging infrastructure by doubling the number of Level 2 and DC Fast Chargers (DCFC) available to the public.

- *Infrastructure*: PDX now has a $1.3 billion plan to modernize the terminal core of the airport, based on serving 35 million passengers annually by about 2035, up from 23 million passengers each year today [49]; Portland's Jetport expecting $312 M upgrade too;

- *Finance*: TNC service fee monthly revenues for the airport has increased from $22,122 in May, 2015 to $318,966 in October, 2017. Unlike other cities, monthly parking revenue has also increased—roughly doubling from October 2010 ($3,645,956) to October 2017 ($6,222,862)—this is over a period of 140% growth in passenger volumes. However, on a parking revenue per passenger basis, parking revenues have declined ~8% over the 24 months after TNCs entered the market at PBX (as shown in results, **Figure 2**).

- *Resilience*: The City of Portland was proactive in collaborating on data with TNCs as related to change in TNC and taxi trips by zip code, with large TNC trip increases seen at the airport and average TNC ride duration being highest by the airport as originating zip code [50].

5. Dallas

- *Mobility*: World's 10th busiest airport in 2015 (64.1 M travelers); 4th busiest US commercial airport, with a projected doubling of future freight by 2040 and passenger travel will increase by >50% too; a key sustainability goal includes optimizing efficiency of fleet operations;

- *Energy*: 26 EV charging stations and goals of decreasing energy/fuel use, to conversion of vehicles as increasing alternative energy portfolio; aims of carbon neutrality; leadership has noted that utility owns/operates the transmission/distribution systems and have not yet embraced the electric vehicle (and are perhaps not sure how it will impact the system); announcements of UberElevate's first electric flying shuttles at both DFW and LAX, offer unique areas for analyses.

- *Infrastructure*: airport encompasses ~18,000 acres, making it second largest in the US in terms of land area; $1.9 billion renovation and expansions including building a 6th terminal; and new considerations emerging when developing parking garages; thinking is changing due to new rail and TNCs—questions on future demand for parking, roadways, fueling infrastructure and energy infrastructure? 'Will we have 1000s of cars parked in garage and as EVs with an extension cord? Is airport ready for new loads of EVs and how will infrastructure be put in place to attract business? A need to understand how advances in transportation/mobility systems impact airport as automated vehicles/buses?'

- *Finance*: airport chief financial officer estimated that the airport lost somewhere between $10 million to $20 million in revenue to Uber and Lyft [51]; airports concerned on capacities to dynamically set TNC service fees to raise revenues; also exploring ways to make parking products attractive (e.g., valet services) rather than empty spaces or high curbside demand with limited supply getting zero revenue, despite potential for willingness to pay for easier access.

- *Resilience*: A desire to build infrastructure that is flexible and can adapt to new demands; to ensure resilience strategy also helps address water-scarce region, droughts, and high water demands (e.g., for DFW central utilities plant cooling towers, irrigation of open space, and gas drilling operations on airport property) via reclaimed water strategy with City of Fort Worth.

These snapshot examples of airports and cities offer a diverse range of the sets of challenges, priorities and trends emerging. An important comparison, in the context of emissions benefits of EV adoption in urban areas and for airports aiming towards ambitious climate action goals, is shown in **Figure 3** of the States of Texas

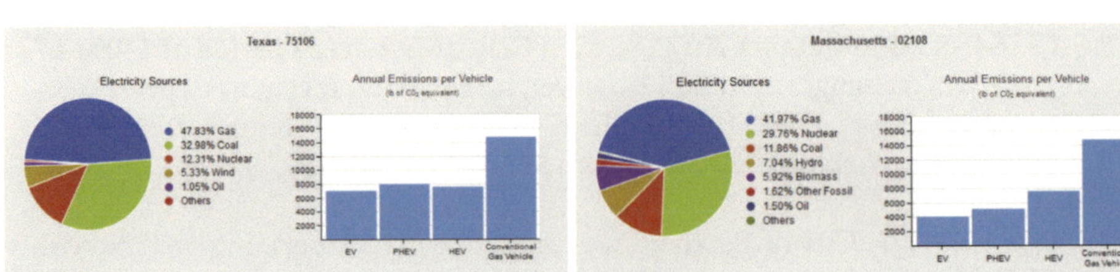

Figure 3.
Texas versus Massachusetts grid mix: associated annual emissions per vehicle (source: To and Linnewiel [52]).

versus Massachusetts, in terms of the overall life-cycle well-to-wheels emissions benefits for EVs, that vary by grid mix. For example, the annual EV emissions is almost double in Texas.

Table 2 offers another literature and data review approach, comparing Pittsburgh, Columbus, Austin, Atlanta and Houston airports, on emerging investments and recent disruptions that may inform future intelligent transportation management, modeling and control in the context of electrification dynamics.

These reviews motivate several questions for on-going research, development, demonstration, and long-term planning, using data to generate understanding on how best to help reduce costs, improve mobility, save energy, modernize infrastructure assets, build resilience and ensure sustainable revenue models. Some of these questions associated with data and assessment methods are described in **Box 1**.

2.3 Exploring impacts of new technology services and extreme events: a case study review

In 2011, when Tropical Storm Irene arrived, all major New York airports were closed. Although not a hurricane, the storm resulted in 5–8 in. of rain, and generated news that certain categories of hurricanes would put JFK International Airport under more than 15 ft. of water. This very substantial risk is known and now very well understood by the airport, metro region, and whole Northeast mega-region after Hurricane Sandy hit in 2013—as this event cost US airlines an estimated US$250 million in revenue due to ~20,000 flight cancelations on the eastern seaboard of the US from DC to Boston. These disruptive events, as well as disruptions associated with new technology services, offers new opportunities to explore the range of impacts to e-mobility, energy/grid and infrastructure impacts. A consequence of not exploring risks and vulnerabilities includes developing a reputation of mishandling disruption. The long-term business impacts from passengers who might make other travel choices could be substantial. Therefore, improved understanding of impacts can inform new EV planning, decisions, to future optimal responses for efficient, affordable EV systems and long-term resilience (e.g., managing risks/recovery).

Approach. Different strands of scholarship and practice provide context on the important impacts relevant to the design and upgrading of airports and urban systems as they transition. This review includes literature on impact assessment, including benefits of and risks to airport and urban transitions and transformations. This then informs a quadrant-based framework (**Figure 2**) that aims to further explore the synergies and trade-offs between these risks and benefits. For example, advances in energy efficient mobility and infrastructure may not necessarily lead to sustainable revenue and finance models for sustainable and resilient upgrades that will be able to account for future stresses, shocks and potential disruptions (that may not yet be identified, similar to airport planning, and investment decision-making prior to the arrival of transportation network companies such as Uber and Lyft). Similarly, growing revenue bases and finance may not necessarily enable moving towards more efficient resource utilization in terms of mobility, energy, or infrastructure systems and services.

This literature review takes two approaches. First, potential impacts are reviewed. Second, review of how shared mobility, electric mobility, and connected and automated vehicles (CAVs) may disrupt airports is reviewed. Progress at both airports and in urban areas is explored, in order to identify knowledge and scenario modeling gaps to be filled and opportunities for energy efficient systems that reduce costs, improve operations, economic prosperity, and resilience.

Airport	Mobility	Energy	Infrastructure	Finance	Resilience
Pittsburgh	~9 M annual passengers and 75 k tons of freight; ~8% growth rate for both from 2016 to 2017	9000 acres of airport leased for natural gas drilling ($50 M signing and ~$16 M/yr. in royalty)	25 years old; $1.1B renovation by 2023; new 4500-space parking garage at a cost of $258 M	Saving $23 M/yr. via upgrades and financing via 20- to 30-year bonds, grants, passenger facility charges, natural gas drilling revenue.	Airport viewed as essential to regional economy/in times of emergency; aging infr/power outages motivate dev of smart transport system and airport microgrid
Columbus	Estimate: ~9 M passengers annually by mid-2020s; Columbus Regional Airport Authority as also a transit authority (plus COTA); city future pilots of AV shuttles	Goal of continuous pursuit of energy conservation, efficiency and innovation for moving both people and cargo—e.g., efficient trucks—test of platooning	$2B for new terminal, parking garage, rental car facility, utility upgrades—addressing overcrowded facilities, 1950s-built airport (Note: an $80 M 2016 renovation)	Revenue records not yet available—parking, car rental, TNC, taxi and ground transportation data still being collected. With 2% of total trips by transit, most revenue loss may be for parking and car rentals	Smart city aware helping to address underdeveloped bus system and minimal public transit network (largest US metro with no passenger rail or light rail), leaving city almost entirely (89%) reliant on private cars for travel
Austin-Bergstrom	13.9 M airport passengers, ('17), 12% increase ('16) as 8 year in a row of growth; 48% less passengers in 1999 (when opened); 500% increase in MetroRail ridership from 2008 to 2014.	Municipally owned utility, Austin Energy, enabling state-of-the-art transport electrification strategy for the airport and city-wide	Two new parking garages, 9-gate expansion, and project on airport boulevard corridor to better accommodate safe, multimodal connectivity	Use of a public-private partnership to build a 1.6 M SF CONRAC rental car facility; 2016 mobility bond; according to 2015 poll, 63% would favor an increase in taxes to construct above ground rail	Focus on airport security and continuity of operations; ASOCS is their web-based operational and communications management system that helps manage and reports all incidents and activities at the airport.
Atlanta	>250,000 passengers daily;	Recent 11 h power outage led to cancelation of over 1000 flights	$6B capital improvement plan: CONRAC rental car center, a fifth runway, int'l terminal	InterContinental Hotel as one key opportunity to diversifying revenue [9]	Reliable, redundant and resilient energy supply for airport-critical facilities via district-scale energy solutions
Houston	53 M passengers annually, with more than 10 million as international travelers	Focused on reduced energy costs (e.g., chiller plant); responding to issues with sensors; and reconfiguring water treatment schedule to reduce pumping	$1.5B expansion and renovation including $244 M for new Terminal C North concourse; airport sits on more than 11,000 acres; airport system as three airports;	Projected 18 month payback through energy infrastructure upgrades and $1 M in annual savings; airport system fund from the city of Houston; FEMA funding of $114 M for overall Houston flood recovery	Airport emergency operations center tested by Hurricane Harvey, along with security of passengers and employees; flights were canceled, roads were flooded, and some passengers were stranded and employees were not able to leave after normal shifts

Table 2.
Comparisons of city airports, unfolding megatrends, and resilience priorities.

Data

- How can the larger intelligent, smart cities and e-mobility research community advance data-driven discovery, planning and decision-making—what EV strategies, enhanced data and modeling can best inform responses to rising airport passenger/GHG levels?
- What strategies will encourage innovation and reinventing of U.S. airport ground transportation across emerging mobility-energy-infrastructure service environments?
- What will be the future of parking revenue, land and infrastructure (re)development, and in what direction to we need to move to harness technology and new services for positive economic/business model outcomes while reducing financial or other risks?

Assessment Methods

- What are key priorities, modernization definitions, and related metrics for reporting?
- What EV mobility input data, models, and information resources can be drawn upon?
- How best to integrate sustainable and resilient infrastructure assessment tools for increasingly shared, electric mobility and related infrastructure at airports/in cities?

Box 1.
Key questions for exploring long-term impacts from airport transformations.

Potential impacts of four mega-trends. First-in-history megatrends, described below, touches on emerging themes of EV technology, market growth, and new consumer demands for better services motivated by economics, time, convenience, socio-demographics, and so on. Cities, their airports and broader transportation systems that are ever-increasingly connected.

From review, impacts can be largely categorized into the megatrends of changes in:

- *Mobility efficiency and costs:* electric vehicle (EV) battery prices of more than $1200 per kWh in 2005 are soon to be as low as $200 per kWh for some suppliers in 2020 [5]; shared mobility and mobility as a service (MaaS) choices are expected to continue dropping in costs as urban and airport markets continue to emerge and capture larger market shares, with more optimal matching of supply and demand, lower costs through 'pooling', and future automation. The current global sharing economy market estimated at over US$250bn (led by Uber) (US$68bn, #1 global ride-sharing); Didi Chuxing (US$50bn, #1 China ride-sharing) and including Airbnb, WeWork and others could soon reach an estimated US$2 trillion globally [53].

- *Energy efficiency and new generation costs enabling distributed energy resources as PV, EV, and storage:* solar photovoltaic prices declined almost 75% from 2010 to 2017 and onshore wind electricity by 25% to $0.06USD/kWh in 2017 [54]. Bids for solar plus storage or wind plus storage are now cost-competitive with almost all other options, leading to new emerging markets for distributed energy resources, especially at airports where microgrids provide operational flexibility, cost savings, and new redundancies in offering reliable EV fueling options for groundside to airside transportation.

- *Infrastructure efficiency and integration:* some of the largest investments in infrastructure today are being made in strengthening, retooling and integrating new technologies and cyber-physical services at airports (and in cities). There is significant potential for maturity in the use of more intelligent, dynamic management and control of energy, mobility, building, to communications infrastructure. This includes improved supply chains and distribution systems for services, to integrated utility/grid/network management, mobility systems integration to achieve future visions [55].

- *Finance/revenues* unique opportunities are emerging to increase revenue and financing (e.g., utilities via transportation electrification; airports via TNC service fees and/or future EV incentives). Attracting new finance, enabling of public-private partnerships, or new business models can sustainably enable more market-based and self-sustaining airport capital improvement, EV infrastructure upgrades, and operational resilience strategies—where airports often are the innovators, cities could be considered early adopters, with other actors as later adopters or laggards, in less mature markets for adapting and learning from new approaches to rate/fee structures within context of new services being adopted in the public 'right of way' [56, 57].

There are more specific details that can be inferred under this framework for the type of impacts associated with these megatrends, yet this articulation provides an area for high-level modeling analysis opportunities on potential risks and benefits. As noted in **Figure 4**, a synergistic effect could take place in the upper right category, where more efficient, cost-effective, and satisfactory services for customers may provide a virtuous cycle with revenue growth associated with continued investments in modernization towards efficient e-mobility, energy and infrastructure with enhanced resilience and vice versa as leading to sustainable revenue growth, finance, and increased economic prosperity/competitiveness of the region.

2.4 Preparing: mobility disruptions of shared, electric, connected to automated systems.

While studies being performed analyzing the impact of shared and electric mobility, to the potential future impacts of CAVs, few have explored impacts directly at airports and/or specifically as focusing EV adoption first on high-mileage, high utilization, and high occupancy assets. Significant opportunities may exist by enabling this emphasis, with focus on data with TNC drivers to site fast-charging infrastructure, or incentives to enable changes associated with EVs waiting in airport queues, to other equity focus areas for airports, cities, and electric utilities to consider—that could emerge by efficiently moving TNC EVs through queues or passengers having use of curb and a more affordable trip, if

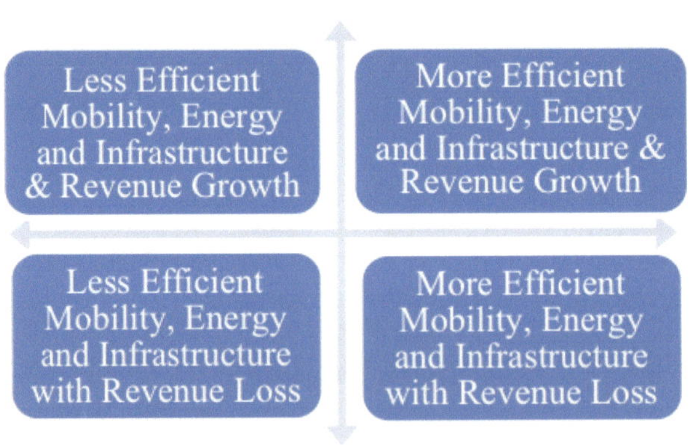

Figure 4.
Mapping impacts as synergies and tradeoffs between goals and financial health.

requesting an EV and/or higher occupancy ride-hail option for their ride in a TNC or public transit vehicle.

With human behavior, future factors to consider and collect data, generate modeling and control approaches on could include examining e-mobility dynamics through perspective of key use cases, e.g.,

- Prioritizing low and moderate income (e.g., accessibility, equity, and potential for a used EV market).
- Prioritizing high-mileage vehicles (e.g., to accelerate decarbonization pathways)—including with transit, ridehailing (TNC), and Taxi—as prior to privately owned vehicles.
- Broader equity contexts in terms of affordable housing, transportation and EVs—e.g., emphasis on multi-unit dwelling (garage orphans that lack charging at home and those who spend over X% of their income on energy and housing).
- Utility planning—with emphasis on grid impacts/management, incentives for diverse customers, and generating new revenues.

Additional considerations could include:

- easier or inhibited travel (e.g., free and/or easier access to curbside or door-to-door travel if higher occupancy, electrified vehicles; or up to $10 service fees added or less convenient parking garage pickup and drop-off zones for single occupancy, and gas vehicles),
- Variations in travel by higher income or working populations (at home or away airports),
- Potential for short-haul air travel as modal shifts to increasingly affordable AV-driven e-mobility,
- Shared-automated vehicle ride-hailing, to empty miles/deadheading, and shifting parking demands.

Introduction of electrified MaaS and CAVs may also result in changes to costs, and financing via per-mileage usage fees rather than prior gas tax revenue. Increases in costs may be due to: deploying new technology in airport vehicle fleets, higher maintenance and repair costs as larger infrastructure footprints emerge, service fees, and demand growth. Reductions in cost may be due to: insurance premiums, less damage due to power outages, or other operational costs. Predicted impacts of shared, e-mobility including vis-à-vis automation continues to have high levels of uncertainty in terms of rates of adoption, willingness and ability to share, and access/egress constraints or enablers for higher levels of energy efficient mobility.

3. Results

The results described below include an emphasis on managing growth in passengers, as well as shifts in TNC service fees and parking revenues—which touch on most aspects of the systems and megatrends described. For simplicity purposes, we zoom in specifically on integrating a few datasets. This includes an emphasis on TNC data inputs and new service fee analyses with potential to inform new mobility and energy efficient mobility system scenarios, case studies for new strategies, and/or improved investments.

For example, based on the trends shown in the latest cross-city airports analysis below, it is expected that many of the airports identified will need to devise new finance models that may include monetizing ground access, for example by introducing a per-person access fee or occupancy-based fee. With high uncertainty as to the future of parking demand decreases and to what threshold will trends hold consistent, there will likely need to be a reconsideration of whether investments in enormous new parking garages/infrastructure, as well as CONRAC facilities are needed. At the same time, this may also be less of a challenge for airports that have historically limited parking due to land and geographic constraints (e.g., Boston Logan Airport).

While many new structures will not be ready until the mid-2020s, it is very possible that with up to a 50% (or even a more conservative 20%) drop in parking demand, some investments may quickly become obsolete. Instead, successful airports may develop scenarios for multimodal ground transportation that can flexibly accommodate a variety of access modes, identify explanatory or predictive factors for new mode choices, and anticipate the different needs of passengers traveling on each of them. In addition, strategies for adaptive investments in new affordable and cost-saving travel and energy systems, that enable higher occupancy, electric mobility systems and more reliable airport travel facilities could also be in high demand. This is demonstrated by adaptive responses such as a recent Boston airport decision to set up designated ride-hailing pickups from a centralized garage site; LAX airport banning Uber, Lyft, or taxi pickups on the curbside and instead requiring passengers to use shuttles to transport from farther away land over to terminals; and the Port Authority of New York New Jersey now defining a pathway to implement occupancy-based pickup and drop-off fees to incentivize less congestion to and from the airport and enable more efficient groundside airport operations.

3.1 Cross-city airport trends based upon latest revenue data integration and analysis

As noted, airports are experiencing increases in air travel, and with an ever faster increase in ride-hailing travel demands to and from airports. The analyses presented below focused on gaining observability into mode choice trends, quantifying changes using airport data (via public information on passengers, revenue data) to transit data (that has been made available initially for both the Denver and Seattle airports) [10, 58] (**Figure 5**).

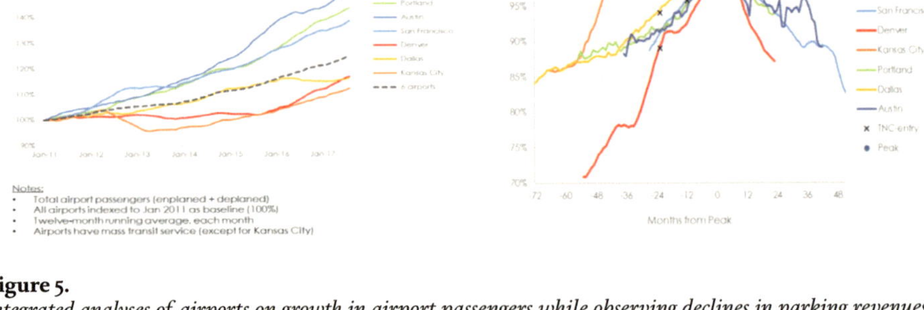

Figure 5.
Integrated analyses of airports on growth in airport passengers while observing declines in parking revenues (similar data has been analyzed indicating similar trends for car rentals).

A refined analysis approach is now emerging focused on transactions per air passenger, shown below for Denver and Seattle. For this specific approach, we gathered monthly transactions from January 2010 to December 2018 with the following data via public requests:

- Airport passengers (enplaned and deplaned) collected by airports
- Ground transportation transactions: parking, car rental, taxis, TNCs (source: airports)
- Transit transactions collected by the corresponding public transportation authority (Sound Transit for Seattle-Tacoma and the Regional Transit District for Denver)

Using transaction data, and based on the number of transactions per mode after ride-hailing introduction, the following are key mode replacement findings:

- Seattle (SEA-TAC) airport: for every 100 new TNC transactions for ground transportation, ~27% replaced transit, 35% replaced parking, 17% replaced car rentals, and 21% replaced taxis
- Similarly, at Denver International (DEN), ride-hailing transactions replaced transit, parking, car rental and taxis at a rate of 34.7, 39.0, 16.6, and 9.7%, respectively.
- Note: Ride-hailing is currently 25–40% cheaper than taxis.

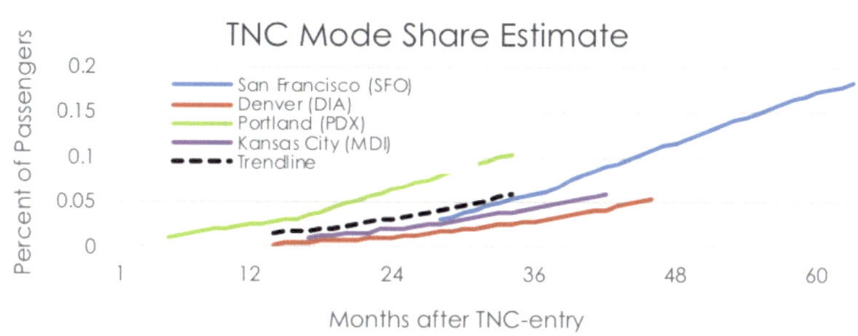

Figure 6.
Integrated analyses of six airports showing how the steady decline in airport parking revenue per passenger is associated with the TNC mode share growth at many airports.

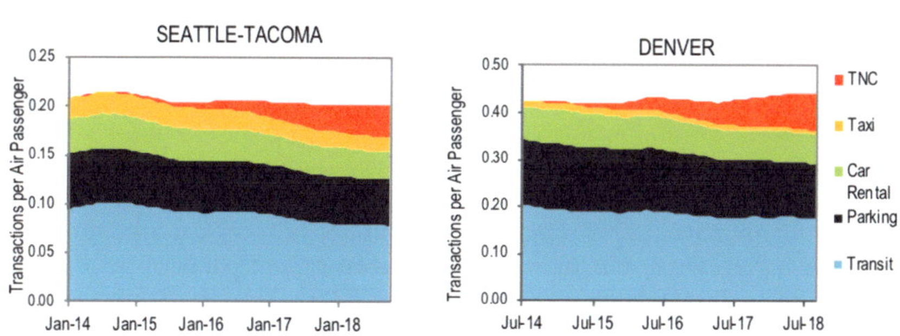

Figure 7.
Transaction data per air passenger and airport mode choice changes.

Figure 6 is a map of fees assessed by airports on TNC companies for ride pickups and/or drop-offs. High charges include Orlando ($5.80), to $5.00 charges in Seattle, San Francisco, Chicago, and Detroit. SFO has been charging this amount since July 1 as per trip (increased from $3.80 to $5 on July 1, 2018, with exceptions made for certain airport access points); in addition, it's worth noting the lack of fees charged by airports in the New York region (**Figure** 7).

4. Discussion: key messages from airports for informing cities

The future of energy-efficient, intelligent mobility systems, will likely include focus on harnessing the potential of electrification of high-mileage, higher-occupancy vehicles across ride-hailing, taxi, to public transit or airport shuttle operations. Thematic areas for future research include:

Exploring human behavior and mobility impacts: What can other areas (e.g., cities, regions) that currently lack TNC- or other private mobility data learn from airports, that may be more data-rich in understanding changes in travel behavior, due to policy and fee-based responses? Will this data inform advanced energy management, modeling and control for hybrid to fully electric vehicles and can ridehailing research inform how connected-automated vehicles are deployed in the contexts of cities and with initial use cases being developed at airports (e.g., EasyMile EZ10 automated-electric shuttle to soon be deployed at Dallas-Fort Worth Airport)?

Exploring new EV infrastructure, technology integration and emerging policies: Can we identify data-driven insights that improve energy-efficiency and financing of multi-modal right-sized electric transportation starting at airports, where incentives for higher vehicle occupancy, electric transit, and infrastructure, coupled with energy (e.g., distributed energy resources, microgrids), and other forms of infrastructure upgrades and modernization could move in more agile ways to accelerate the electrification of transportation, as starting with high-mileage TNCs?

Future of mode choice modeling, scenarios and analyses: Can research explore the modification of airport mode choice and energy modeling by identifying key factors shaping personal versus business traveler, to airport employee, individual versus family travel decisions? What is the appropriate balance to strike in the right-sizing of multiple systems—e.g., parking facilities, curbs, vehicle fleets, to distributed energy resources—and for down-sizing, decommissioning or repurposing certain land uses and infrastructure assets? What are the strategic capacities for changes at airports versus within cities and metro regions? Additional exploring via analyses of new mobility shifts, and building on additional studies will be critical [10]. How different will impacts, priorities, to models of transactions, revenue and financing strategies look across airports and between airports and their respective cities?

5. Conclusions

With airports and cities hosting significant critical infrastructure systems and operations, and air travel today representing over 9% of total U.S. transportation energy use [59], what will the future hold as energy impacts from new mode choices and utility demands for e-mobility? With $3.5 billion in parking and ground transportation fees representing 41% of the $8.5 billion in U.S. airport revenue not related to airlines, according to the Federal Aviation Administration, what will be the future of revenues as TNC fees to utility revenues from e-mobility business models? What is the long-term viability of a nearly US$6B airport parking/car rental industry? Case study airports and their adoption of strategies in increasingly mature markets for on-demand, high-mileage mobility services are expected to offer insights into future opportunities for electric vehicles, shared mobility electrification, finance, and infrastructure (re)development. Future research could focus on integrating public and private community interests for advancing technologies and modernization strategies that have positive real-world implications and upscaling potential for future city systems integration, accelerated planning and decision-making.

Acknowledgements

This study was supported by the U.S. Department of Energy's National Renewable Energy Laboratory under a U.S. Department of Energy (DOE) Systems and Modeling for Accelerated Research in Transportation (SMART) Mobility Laboratory Consortium Program.

Author details

Josh Sperling* and Alejandro Henao
National Renewable Energy Laboratory, USA

*Address all correspondence to: joshua.sperling@nrel.gov

References

[1] Jenn A. Emissions Benefits of Electric Vehicles in Uber and Lyft Services. 2019. Accessed: https://escholarship.org/uc/item/15s1h1kn

[2] Pucher J, Buehler R, editors. City Cycling. MIT Press; 2012. JSTOR. Available at: www.jstor.org/stable/j.ctt5hhjxr

[3] Shaheen S, Martin E, Bansal A. Peer-to-peer (P2P) carsharing: Understanding early markets, social dynamics, and Behavioral impacts. 2018. DOI: 10.7922/G2FN14BD. Retrieved from: https://escholarship.org/uc/item/7s8207tb

[4] Henao A. Impacts of ridesourcing-Lyft and Uber-on transportation including VMT, mode replacement, parking, and travel behavior [doctoral dissertation]. University of Colorado at Denver; 2017

[5] Sperling D. Three Revolutions: Steering Automated, Shared, and Electric Vehicles to a Better Future. Island Press; 2018

[6] Clewlow R. The climate impacts of high-speed rail and air transportation: A global comparative analysis [PhD thesis]. Massachusetts Institute of Technology, Engineering Systems Division; 2012

[7] Federal Aviation Administration (FAA). Capacity Needs in the National Airspace System: An Analysis of Airports and Metropolitan Area Demand and Operational Capacity in the Future. The MITRE Corporation, Federal Aviation Administration, U.S. Dept. of Transportation; 2007

[8] Jacobs Consultancy. Regional Aviation Strategic Plan. San Diego County Regional Airport Authority; 2011

[9] Tretheway M. Addressing Uncertainty about Future Airport Activity Levels in Airport Decision Making. ACRP 03-22 Project Description. 2010

[10] Henao A, Josh S, Venu G, Yi H, Stanley Y. Airport analyses informing new mobility shifts: Opportunities to adapt energy-efficient mobility services and infrastructure. Preprint. NREL/CP-5400-71036. Golden, CO; 2018. Available at: www.nrel.gov/docs/fy18osti/71036.pdf

[11] Coogan et al. Ground Access to Major Airports by Public Transportation. ACRP Report 4. 2008. Available at: https://www.aci-na.org/static/entransit/acrp_access_to_airports.pdf

[12] Mandle P, Box S. Transportation Network Companies: Challenges and Opportunities for Airport Operators. Airport Cooperative Research Program: Synthesis 84. National Academies Press; 2017

[13] Los Angeles Times. LAX to end curbside pickup and dropoff by Uber and Lyft. 2019. Available at: https://www.latimes.com/california/story/2019-10-03/lax-to-end-curbside-pickup-uber-lyft

[14] Hou Y, Garikapati V, Sperling J, Henao A, Young SE. A deep learning approach for transportation network companies trip-demand prediction considering spatial-temporal features. In: Presented at the Transportation Research Board (TRB) Annual Meeting 2019, 13-17 January; Washington, DC. 2019

[15] Romero L, Patricia W, Alana S, Joshua M, Clark Z-S, Daniel B, et al. 2019. Urban Electrification: Knowledge Pathway Toward an Integrated Research and Development Agenda. Available at SSRN: https://ssrn.

com/abstract=3440283 or 10.2139/ssrn.3440283 [Accessed: 20 August 2019]

[16] American Society of Civil Engineers (ASCE). Infrastructure Report Card: A Comprehensive Assessment of America's Infrastructure. 2017. Available at: www.infrastructurereportcard.org/wp-content/uploads/2016/10/2017-Infrastructure-Report-Card.pdf [Accessed: July 2018]

[17] Airports Council International-North America (ACI-NA). U.S. Airport Infrastructure Needs Through 2019 Total $75.7 Billion. 2016. Available at: www.aci-na.org/newsroom/press-releases/us-airport-infrastructure-needs-through-2019-total-757-billion [Accessed: July 2018]

[18] ACI-NA. Airport infrastructure needs: 2017-2021. 2017. Available at: https://www.aci-na.org/sites/default/files/2017infrastructureneedsstudy-web.pdf [Accessed: July 2018]

[19] NYPA. New Statewide Initiative Spur Widespread Adoption of Electric Vehicles and Charging Infrastructure. 2018. Available at: https://www.nypa.gov/news/press-releases/2018/20181119-evolve

[20] Intelligent Transport. Twenty electric buses introduced for airside use at LAX. 2018. Available at: https://www.intelligenttransport.com/transport-news/67287/twenty-electric-buses-airside-use-lax/

[21] National Academies of Sciences, Engineering, and Medicine. Renewable Energy as an Airport Revenue Source. Washington, DC: The National Academies Press; 2015. DOI: 10.17226/22139. Available at: https://www.nap.edu/catalog/22139/renewable-energy-as-an-airport-revenue-source

[22] FAA. NextGen. 2018. Available at: https://www.faa.gov/nextgen/what_is_nextgen/ [Accessed: July 2018]

[23] Department of Transportation (DOT). Round Two: Seven Finalists Create Plans to Implement Their Visions. 2017. Available at: https://www.transportation.gov/smartcity/7-finalists-cities [Accessed: July 2018]

[24] IEA. Global EV Outlook 2019. Paris: IEA; 2019. Available at: www.iea.org/publications/reports/globalevoutlook2019/

[25] Alba S, Manana M. Energy research in airports: A review. MDPI Energies. 2016;**9**:349

[26] Marcotullio PJ, Sarzynski A, Sperling J, Chavez A, Estiri H, Pathak M, et al. Chapter 12: Energy transformation in cities. In: Rosenzweig C, Solecki W, Romero-Lankao P, Mehrotra S, Dhakal S, Ibrahim SA, editors. Cambridge, UK; New York, NY: Cambridge University Press, Columbia University; 2018

[27] Cutter S, Solecki W, et al. Urban systems, infrastructure, and vulnerability. In: Chapter 11—Urban. Third NCA (Draft v. 22 November 2013). 2013. Available at: www.globalchange.gov/sites/globalchange/files/Ch_11_Urban_GovtReviewDraft_Nov_22_2013_clean.pdf [Accessed: July 2018]

[28] ACI-NA. Airport financing. 2018. Available at: https://www.aci-na.org/content/airport-financing [Accessed: July 2018]

[29] Floater G, Dowling D, Chan D, Ulterino M, Braunstein J, McMinn T. Financingtheurbantransition:Policymakers' summary. Coalition for Urban Transitions. London/Washington, DC. Available at: http://newclimateeconomy.net/content/cities-working-papers

[30] San Francisco County Transportation Authority (SFCTA). Emerging Mobility Evaluation Report: Evaluating Emerging Mobility Services and Technology in San Francisco. 2018.

Available at: https://www.sfmta.com/sites/default/files/reports-and-documents/2018/05/6-7-18_cac_item_7_emerging_mobility_evaluation_report.pdf

[31] Steele. Ace parking says Uber, Lyft have cut business up to 50% in some venues. San Diego Tribune. Business/Growth & Development. February 22, 2018

[32] San Francisco Airport (SFO). Zero Net Energy. 2018. Available at: https://www.flysfo.com/environment/zero-net-energy.Airport Development Plan Utility: Projects. http://media.flysfo.com.s3.amazonaws.com/default/about/Utility_Projects.pdf [Accessed: July 2018]

[33] Smith D. Why airports are perfect microgrid and resilient power candidates. 2018. Available at: www.solarpowerworldonline.com/2018/07/why-airports-are-perfect-microgrid-and-resilient-power-candidates/

[34] McGinnis C. Uber, Lyft fees to increase at SFO in July. 2018. Available at: https://www.sfgate.com/travel/article/Uber-Lyft-fees-to-increase-at-SFO-13026867.php [Accessed: July 2018]

[35] Murray J. $78.9 million expansion of DIA's underground concourse train system approved by Denver council. 2018. Available at: www.denverpost.com/2018/07/23/dia-underground-train-expansion/ [Accessed: July 2018]

[36] National Academies of Sciences, Engineering, and Medicine. Airport Revenue Diversification. Washington, DC: The National Academies Press; 2010. DOI: 10.17226/14386

[37] Panasonic, Xcel Energy, Younicos. A Portfolio Microgrid in Denver, Colorado: How a Multi-Use Battery Energy Storage System Provides Grid and Customer Services through a Public-Private Partnership. January 2017. Available at: https://www.younicos.com/wp-content/uploads/2017/02/201702-Microgrid-White-Paper.pdf

[38] DIA. Energy Master Plan. 2018. Available at: http://business.flydenver.com/bizpdf/DIAID_2176.pdf

[39] Svaldi A. DIA does not track Uber and Lyft rides, opening risk of losing revenue, Denver Auditor Warns. 2018. Available at: https://www.denverpost.com/2018/02/15/dia-uber-lyft-revenue-audit/ [Accessed: July 2018]

[40] KCI. Traffic Stats. 2018. Available at: www.flykci.com/newsroom/statistical-and-financial-info/traffic-statistics/

[41] KCI. Annual Passengers at Kansas City International Airport Exceed 11 Million in 2016. 2017. Available at: www.flykci.com/newsroom/news-releases/kci-passengers-exceed-11-million-in-2016/

[42] VanLoh. Aviation in Kansas City. Director of Aviation. 2013. Available at: https://data.kcmo.org/download/96w8-am59/application%2Fpdf [Accessed: July 2018]

[43] KCI. Terminal Area Master Plan. City of Kansas City: Aviation Department; 2015. Available at: https://www.flykci.com/media/1446/mci-terminal-area-master-plan-april-2015-final.pdf

[44] Unison Consulting. Economic Impact of Kansas City International Airport. 2015. Available at: http://www.flykci.com/media/1454/kci-fy-2014-economic-impact_final_072415.pdf

[45] City of Kansas City. DOT smart city challenge. Beyond Traffic: The Vision for the Kansas City Smart City challenge—Part 1. Vision Narrative. 2016. Available at: https://cms.dot.gov/sites/dot.gov/files/docs/Kansas%20

City%20Vision%20Narrative.pdf [Accessed: July 2018]

[46] City of Portland. 2017 EV strategy. 2017. Available at: https://www.portlandoregon.gov/bps/article/619275

[47] Forth. Transportation electrification strategies for electric utilities. 2019. Available at: https://forthmobility.org/storage/app/media/Documents/Forth-Utility-White-Paper.pdf

[48] Portland International Airport. Airport futures: Charting a future for PDX. In: Sustainability Report. 2010. Available at: airportsustainability.org/LibraryDownload/document/4 [Accessed July 2018]

[49] Cheney C. The Portland airport's billion-dollar bet. Available at: www.bizjournals.com/portland/news/2017/03/23/the-portland-airports-billion-dollar-bet.html [Accessed: July 2018]

[50] Novick S, Treat L. Portland's Private For-hire Transportation Market: Summary Report of the PFHT Innovation Pilot Program. Portland Bureau of Transportation; 2015

[51] Williamson R. Airports, facing growth of ride-hailing apps, enhance parking technology. 2018. Available at: https://www.bondbuyer.com/news/airports-guard-parking-revenues-against-ride-hailing-services

[52] To K, Linnewiel R. Electric vehicles: What, why and how? Dallas Fort-Worth Clean Cities. Available at: https://www.nctcog.org/nctcg/media/Transportation/DocsMaps/About/Educate/Presentations/2015%20Presentations/2015-10-23_TerrellISD.pdf

[53] Durden T. A primer on the "global sharing economy" in 20 charts. 2017. Available at: www.zerohedge.com/news/2017-07-24/primer-global-sharing-economy-20-charts [Accessed: April 2017]

[54] IRENA. Renewable Power Generation Costs in 2017. Abu Dhabi: International Renewable Energy Agency; 2018

[55] Bouton S, Dearborn J, Sergienko Y, Woetzel J. Unlocking the future: The keys to making cities great. 2015. Available at: https://www.mckinsey.com/industries/capital-projects-and-infrastructure/our-insights/unlocking-the-future-the-keys-to-making-cities-great

[56] Florida R. The Rise of the Creative Class Revisited. New York, NY: Basic Books; 2012

[57] King S, Puentes R. Taxing New Mobility Services: What's Right? What's Next? Eno Brief. 2018. Available at: https://www.enotrans.org/wp-content/uploads/2018/07/Eno_Brief_Taxing_New_Mobility_Services.pdf?x43122 [Accessed: July 2018]

[58] Henao A, Sperling J, Young S. Ground Transportation at Airports: Ride-hailing Uptake and Travel Shifts to Test Mode Choice Modeling Assumptions. In: 2020 Transportation Review Board Annual Meeting. 2019. Accepted

[59] U.S. Energy Information Administration. International Transportation Energy Demand Determinants (ITEDD-2015) model estimates. 2016. Accessed: https://www.eia.gov/todayinenergy/detail.php?id=23832

Chapter 5

Robust Feedback Linearization Approach for Fuel-Optimal Oriented Control of Turbocharged Spark-Ignition Engines

Anh-Tu Nguyen, Thierry-Marie Guerra and Jimmy Lauber

Abstract

This chapter proposes a new control approach for the turbocharged air system of a gasoline engine. To simplify the control implementation task, static lookup tables (LUTs) of engine data are used to estimate the engine variables in place of complex dynamical observer and/or estimators. The nonlinear control design is based on the concept of robust feedback linearization which can account for the modeling uncertainty and the estimation errors induced by the use of engine lookup tables. The control feedback gain can be effectively computed from a convex optimization problem. Two control strategies have been investigated for this complex system: drivability optimization and fuel reduction. The effectiveness of the proposed control approach is clearly demonstrated with an advanced engine simulator.

Keywords: turbocharged gasoline engine, engine control, robust control, feedback linearization, linear matrix inequality

1. Introduction

The control of turbocharged air system of spark-ignition (SI) engines is known as a challenging issue in automotive industry. It is complex and costly to develop and implement a new control strategy within industrial context since it may change the available software in series [1]. The novel control strategies, generally needed when some new technologies are introduced, have to justify its relevant advantages with respect to the actual versions. At the same time, they have to satisfy several stringent constraints such as control performance/robustness, calibration complexity, and software consistency. Therefore, conventional control approaches are still largely adopted by automakers. These control strategies consist in combining the gain-scheduling PID control with static feedforward lookup table (LUT) control [2]. This results in an easy-to-implement control scheme for the engine control unit (ECU). However, such a conventional control strategy remains some inherent drawbacks. First, using gain-scheduling PID control technique and static feedforward LUTs, each engine operating point needs to be defined, leading to heavy calibration efforts. In addition, it is not always clear to define an engine operating point, in particular for complex air system with multiple air actuators [1].

Second, the trade-off between performance and robustness is not easy to achieve for a wide operating range of automotive engines. Therefore, conventional control strategies may not be appropriate to cope with new engine generations for which many novel technologies have been introduced to meet more and more stringent legislation constraints. Model-based control approaches seem to be a promising solution to overcome these drawbacks.

Since turbochargers are key components in downsizing and supercharging technology, many works have been recently devoted to the turbocharged engine control. A large number of advanced model-based control technique have been studied in the literature, e.g., gain-scheduling PID control [3, 4], H_∞ control [5], gain-scheduling H_∞ control [6], sliding mode control [7], predictive control [8], etc. These control techniques are based on engine model linearization to apply linear control theory. Hence, the calibration efforts are expensive and the aforementioned drawbacks still remain. Nonlinear control seems to be more relevant for this complex nonlinear system. Most of the efforts have been devoted to diesel engine control [9–11], and only some few works have focused on SI engine control. In [1], the authors proposed an interesting approach based on flatness property of the system combining feedback linearization and constrained motion planning to meet the predefined closed-loop specifications. However, due to the robustness issue with respect to the modeling uncertainty, this control approach requires a refined control-based engine model to provide a satisfactory control performance. To avoid this drawback, many robust nonlinear control approaches have been proposed for turbocharged engine control, for instance, fuzzy sliding mode control [12], double closed-loop nonlinear control [13], nonlinear model predictive control [14], and so forth. However, for most of the existing control approaches, it is not easy to take into account the *fuel-optimal strategy* [15] in the control design when considering the whole system. To get rid of this difficulty, a novel control strategy based on switching Takagi-Sugeno fuzzy model has been proposed in switching control [16–18]. Although this powerful nonlinear control approach provides satisfactory closed-loop performance, it may look complex from the industrial point of view. In this chapter, we propose a new control design based on feedback linearization for the turbocharged air system which is much simpler (in the sense of real-time implementation) and can achieve practically a similar level of performance as in [19]. To the best of our knowledge, this is the second nonlinear multi-input multi-output (MIMO) control approach that can guarantee the stability of the whole closed-loop turbocharged air system while taking into account the *fuel-optimal strategy* after [20]. Furthermore, the proposed control approach allows reducing the costly automotive sensors and/or observers/estimators design tasks by exploiting the maximum possible available offline information. The idea is to estimate all variables needed for control design by using piecewise multiaffine (PMA) modeling [21, 22], represented in the form of static LUTs issued from the data of the test bench. The effectiveness of the proposed control strategy is illustrated through extensive AMESim/Simulink co-simulations with a high-fidelity AMESim engine model.

The chapter is organized as follows. Section 2 reviews some basis on feedback linearization. In Section 3, a new robust control design based on this technique is proposed in some detail. Section 4 is devoted to the control problem of a turbocharged air system of a SI engine. To this end, a brief description of this system is first recalled. Besides a conventional MIMO control approach, a novel idea is also proposed to take into account the strategy for minimizing the engine pumping losses in the control design. Then, simulation results are presented to show the effectiveness of our proposed method. Finally, some concluding remarks are given in Section 5.

2. Feedback linearization control

Feedback linearization provides a systematic control design procedure for nonlinear systems. The main idea is to algebraically transform nonlinear system dynamics into a (fully or partly) linear one so that the linear control techniques can be applied [23, 24]. However, it is well known that this technique is based on the principle of exact nonlinearity cancelation. Hence, it requires high-fidelity control-based models [25]. This is directly related to the closed-loop robustness property with respect to model uncertainties. To this end, a new robust design dealing with model uncertainties/perturbations will be proposed. Compared to some other existing results on robust feedback linearization [24, 26, 27], the proposed method not only is simple and constructive but also maximizes the robustness bound of the closed-loop system through a linear matrix inequality (LMI) optimization problem [28]. Furthermore, this method may be applied to a large class of nonlinear systems which are input–output linearizable and possess stable internal dynamics.

For engine control purposes, we consider the following input-output linearization for MIMO nonlinear systems:

$$\begin{cases} \dot{x}(t) = f(x) + \sum_{i=1}^{m} g_i(x) u_i(t) \\ y(t) = h(x) \triangleq [h_1(x), \dots, h_m(x)]^T \end{cases} \tag{1}$$

where $x(t) \in \mathbb{R}^n$ is the system state, $u(t) \in \mathbb{R}^m$ is the control input, and $y(t) \in \mathbb{R}^m$ is the measured output. The matrix functions $f(x)$, $g(x)$, and $h(x)$ are assumed to be sufficiently smooth in a domain $D \subset \mathbb{R}^n$. For simplicity, the time dependence of the variables is omitted when convenient.

The feedback linearization control law of the system (1) is given by

$$u(t) = \begin{bmatrix} L_{g_1} L_f^{\rho_1 - 1} h_1(x) & \dots & L_{g_m} L_f^{\rho_1 - 1} h_1(x) \\ \dots & \dots & \dots \\ L_{g_1} L_f^{\rho_m - 1} h_m(x) & \dots & L_{g_m} L_f^{\rho_m - 1} h_m(x) \end{bmatrix}^{-1} \left(\begin{bmatrix} v_1(t) \\ \vdots \\ v_m(t) \end{bmatrix} - \begin{bmatrix} L_f^{\rho_1} h_1(x) \\ \vdots \\ L_f^{\rho_m} h_m(x) \end{bmatrix} \right)$$
$$\triangleq J^{-1}(x)(v(t) - l(x)) \tag{2}$$

where $[\rho_1 \dots \rho_m]^T$ is the vector of relative degree and v is a vector of new manipulated inputs. The Lie derivatives $L_f^{\rho_i} h_i(x)$ and $L_{g_i} L_f^{\rho_i - 1} h_i(x)$ of the scalar functions $h_i(x)$, $i = 1, \dots, m$, are computed as shown in [25] and [24]. Note that the control law (2) is well defined in the domain $D \subset \mathbb{R}^n$ if the *decoupling matrix* $J(x)$ is non-singular at every point $x_0 \in D \subset \mathbb{R}^n$. The new input vector $v(t)$ can be designed with any linear control technique. The relative degree of the whole system (1) in this case is defined as

$$\rho = \sum_{k=1}^{m} \rho_k \tag{3}$$

Depending on the value of the relative degree ρ, three following cases are considered. First, if $\rho = n$, then the nonlinear system (1) is fully feedback linearizable. Second, if $\rho < n$, then the nonlinear system (1) is partially feedback linearizable. In this case, there are some internal dynamics of order $(n - \rho)$.

For tracking control, these dynamics must be guaranteed to be internally stable. Third, if ρ does not exist on the domain $D \subset \mathbb{R}^n$, then the input-output linearization technique is not applicable. In this case, a virtual output $\tilde{y}(t) = \tilde{h}(x)$ may be introduced such that the new system becomes feedback linearizable [25]. The linearized system for the two first cases can be represented under the following *normal form* [23]:

$$\begin{cases} \dot{\xi}(t) = A\xi(t) + Bv(t) \\ y(t) = C\xi(t) \\ \dot{\omega}(t) = f_0(z(t), v(t)) \end{cases} \tag{4}$$

with $z(t) \triangleq [\xi(t), \omega(t)]^T$, where $\xi(t) \in \mathbb{R}^\rho$ and $\omega(t) \in \mathbb{R}^{n-\rho}$ are obtained with a suitable change of coordinates $z(t) = T(x(t)) \triangleq [T_1(x(t)), T_2(x(t))]^T$. The triplet (A, B, C) is in Brunovsky block canonical form. The system $\dot{\omega}(t) = f_0(z(t), v(t))$ characterizes the internal dynamics [23]. Note that if this system is input-to-state stable, then the origin of system (4) is globally asymptotically stable [24].

3. LMI-based robust control design

Modeling errors are unavoidable in real-world applications, especially when using PMA approximation [22]. Thus, a robust design is necessary to robustify the feedback linearization control scheme. This section provides a new robust control approach to deal with this major practical issue. For convenience, the feedback linearization control law (2) is rewritten as

$$u(x) = \alpha(x) + \beta(x)v(t) = \alpha(x) - \beta(x)K\xi(t) = \alpha(x) - \beta(x)KT_1(x) \tag{5}$$

where K is the control gain of the new linearizing controller. The terms $\alpha(x)$ and $\beta(x)$ are directly derived from (2). Due to modeling uncertainty, the real implemented feedback control law can be represented in the form

$$u(x) = \tilde{\alpha}(x) - \tilde{\beta}(x)K\tilde{T}_1(x) \tag{6}$$

where $\tilde{\alpha}(x)$, $\tilde{\beta}(x)$, and $\tilde{T}_1(x)$ are, respectively, the approximations of $\alpha(x)$, $\beta(x)$, and $T_1(x)$. Then, the closed-loop system (4) can be rewritten as

$$\begin{cases} \dot{\xi}(t) = (A - BK)\xi(t) + B\Delta(z) \\ \dot{\omega}(t) = f_0(z(t), v(t)) \end{cases} \tag{7}$$

where

$$\Delta(z) = \beta^{-1}(x)\{\tilde{\alpha}(x) - \alpha(x) + [\beta(x) - \tilde{\beta}(x)]KT_1(x) + \tilde{\beta}(x)K[T_1(x) - \tilde{T}_1(x)]\}|_{x=T^{-1}(z)} \tag{8}$$

The uncertain term $\Delta(z)$ is viewed as a perturbation of the nominal system $\dot{\xi}(t) = (A - BK)\xi(t)$. Assume that the internal dynamics is input-to-state stable. Then, the stability of the system

$$\dot{\xi}(t) = (A - BK)\xi(t) + B\Delta(z) \tag{9}$$

with respect to the uncertain term $\Delta(z)$ is studied. To this end, we assume that the uncertain term $\Delta(z)$ satisfies the following quadratic inequality [29]:

$$\Delta^T(z)\Delta(z) \le \delta^2 \xi^T(t) H^T H \xi(t)\big|_{z=[\xi,\omega]} \tag{10}$$

where $\delta > 0$ is a bounding parameter and the matrix $H \in \mathbb{R}^{l\times\rho}$, characterizing the system uncertainties [19], is constant for a certain integer l. Inequality (10) can be rewritten as

$$\begin{bmatrix} \xi(t) \\ \Delta(z) \end{bmatrix}^T \begin{bmatrix} -\delta^2 H^T H & 0 \\ 0 & I \end{bmatrix} \begin{bmatrix} \xi(t) \\ \Delta(z) \end{bmatrix} \le 0 \tag{11}$$

where I denotes identity matrix of appropriate dimension.

Consider the Lyapunov function candidate $V(\xi(t)) = \xi^T(t)P\xi(t)$, where $P \in \mathbb{R}^{\rho\times\rho}$, $P = P^T > 0$. The time derivative of $V(\xi)$ along the trajectory of (9) is given by

$$\dot{V}(\xi(t)) = \xi^T(t)\Big((A-BK)^T P + P(A-BK)\Big)\xi(t) + \Delta^T(z)P\xi(t) + \xi^T(t)P\Delta(z) \tag{12}$$

If $\dot{V}(\xi(t))$ is negative definite, then this system is robustly stable. This condition is equivalent to

$$\begin{bmatrix} \xi(t) \\ \Delta(z) \end{bmatrix}^T \begin{bmatrix} (A-BK)^T P + P(A-BK) & P \\ P & 0 \end{bmatrix} \begin{bmatrix} \xi(t) \\ \Delta(z) \end{bmatrix} < 0 \tag{13}$$

for all $\xi(t)$ and $\Delta(z)$ satisfying (11). By the S-procedure [28], condition (13) holds if and only if there exists a scalar $\tau > 0$ such that

$$\begin{bmatrix} (A-BK)^T P + P(A-BK) + \tau\delta^2 H^T H & P \\ P & -\tau I \end{bmatrix} < 0 \tag{14}$$

Pre- and post-multiplying (14) with the matrix $diag\left[\tau P^{-1}, I\right]$ and then using the change of variable $Y = \tau P^{-1} > 0$, condition (14) is equivalent to

$$\begin{bmatrix} (A-BK)Y + Y(A-BK)^T + \delta^2 Y H^T H Y & I \\ I & -I \end{bmatrix} < 0 \tag{15}$$

By Schur complement lemma [28], the condition (15) is equivalent to

$$\begin{bmatrix} (A-BK)Y + Y(A-BK)^T & I & YH^T \\ I & -I & 0 \\ HY & 0 & -\gamma I \end{bmatrix} < 0 \tag{16}$$

where $\gamma \triangleq 1/\delta^2$. Using the change of variable $L \triangleq KY$, the control design can be formulated as an LMI problem in Y, L, and γ as follows:

$$\begin{bmatrix} AY + YA^T - BL - L^T B^T & I & YH^T \\ I & -I & 0 \\ HY & 0 & -\gamma I \end{bmatrix} < 0 \tag{17}$$

To prevent the unacceptably large control feedback gains for practical applications, the amplitude of the entries of K should be constrained in the optimization problem. To this end, the following LMIs can be included:

$$\begin{bmatrix} -\kappa_L I & L^T \\ L & -I \end{bmatrix} < 0, \quad \kappa_L > 0$$
$$\begin{bmatrix} Y & I \\ I & \kappa_Y I \end{bmatrix} > 0, \quad \kappa_Y > 0. \tag{18}$$

Note that condition (18) implies $K^T K < \kappa_L \kappa_Y^2 I$ (see [29]). Moreover, to guarantee some prescribed robustness bound $\overline{\delta}$, the following LMI conditions can be also included:

$$\gamma - 1/\overline{\delta}^2 < 0 \tag{19}$$

The above development can be summarized in the following.

Theorem 1. Given a positive scalar $\overline{\delta}$. If there exist matrices $Y > 0$, L, positive scalars γ, κ_L, κ_Y such that the following LMI optimization problem is feasible:

$$\text{minimize} \quad \lambda_1 \gamma + \lambda_2 \kappa_L + \lambda_3 \kappa_Y \tag{20}$$
$$\text{subject to LMI conditions (17)–(19).}$$

Then, the closed-loop system (9) is robustly stable, and the state feedback control law is defined as $u(t) = -K\xi(t)$ where $K \triangleq LY^{-1}$.

The weighting factors λ_1, λ_2, and λ_3 are chosen according to the desired trade-off between the guaranteed robustness bound $\overline{\delta}$ and the size of the stabilizing gain matrix K. The LMI optimization problem can be effectively solved with numerical toolboxes (e.g., [30, 31]).

4. Application to turbocharged SI air system control

The turbocharged air system of a SI engine is illustrated in **Figure 1**. The nomenclature related to the studied system is shown in **Table 1**.

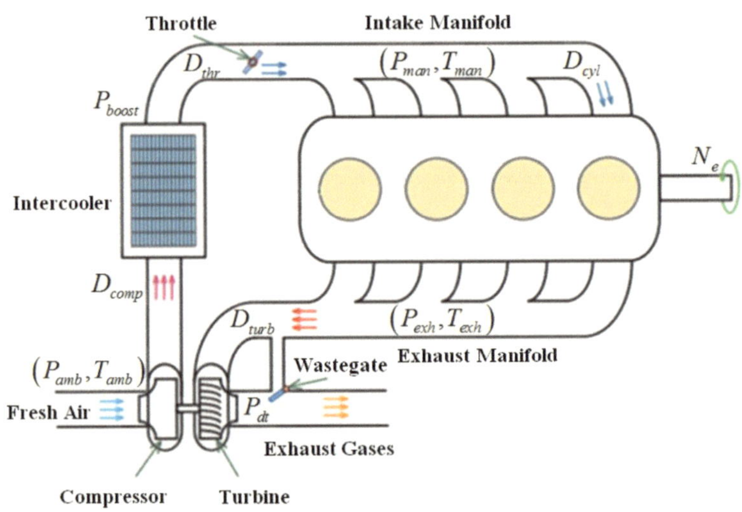

Figure 1.
Schematic of a turbocharged spark-ignition engine.

Symbol	Quantity	Unit	Symbol	Quantity	Unit
Π_{thr}	Throttle pressure ratio	—	D_{cyl}	Cylinder mass airflow	kg/s
Π_{wg}	Wastegate pressure ratio	—	D_{fuel}	Fuel injected flow	kg/s
Π_{comp}	Compressor pressure ratio	—	V_{exh}	Exhaust manifold volume	m^3
Π_{turb}	Turbine pressure ratio	—	V_{man}	Intake manifold volume	m^3
P_{boost}	Boost pressure	Pa	V_{cyl}	Cylinder volume	m^3
P_{man}	Intake pressure	Pa	N_e	Engine speed	rpm
P_{exh}	Exhaust pressure	Pa	$\mathbb{P}_{comp}$	Compressor power	W
P_{dt}	Turbine pressure	Pa	N_{tc}	Turbocharger speed	rpm
P_{amb}	Atmospheric pressure	Pa	$\mathbb{P}_{turb}$	Turbine power	W
T_{amb}	Atmospheric temperature	°K	η_{comp}	Compressor isentropic efficiency	—
T_{man}	Intake manifold temperature	°K	η_{turb}	Turbine isentropic efficiency	—
T_{exh}	Exhaust manifold temperature	°K	η_{vol}	Engine volumetric efficiency	—
D_{thr}	Throttle mass airflow	kg/s	λ_s	Stoichiometric air/fuel ratio	—
D_{wg}	Wastegate mass airflow	kg/s	γ	Ratio of specific heats	—
D_{comp}	Compressor mass airflow	kg/s	R	Ideal gas constant	J/kg/°K
D_{turb}	Turbine mass airflow	kg/s	C_p	Specific heats at constant pressure	J/kg/°K

Table 1.
Notations of turbocharged air system of a SI engine.

4.1 Description of turbocharged air system

Hereafter, a brief description of the air system of a turbocharged SI engine is recalled (see [17, 20, 32] for more details). The model was built with the real data of a four-cylinder turbocharged SI engine from Renault Company (see [19] and [18] for more details). The system dynamics is composed of the three following main parts. First, the intake pressure dynamics is given by

$$\frac{dP_{man}}{dt} = -\eta_{vol}\frac{V_{cyl}}{V_{man}}\frac{N_e}{30}P_{man} + \frac{P_{boost}\sqrt{RT_{man}}}{V_{man}}\Phi_{thr}\left(\Pi^*_{thr}\right)u_{thr} \tag{21}$$

where $\Phi\left(\Pi^*_{thr}\right) = \sqrt{\frac{2\gamma}{\gamma-1}\left(\Pi^{*\,\frac{2}{\gamma}}_{thr} - \Pi^{*\,\frac{\gamma+1}{\gamma}}_{thr}\right)}$ with $\Pi^*_{thr} \triangleq \max\left(\frac{P_{man}}{P_{boost}}, \left(\frac{2}{\gamma+1}\right)^{\frac{\gamma}{\gamma-1}}\right)$ and the volumetric efficiency $\eta_{vol} \triangleq \mathrm{LUT}_{\eta_{vol}}(N_e, P_{man})$ is given by LUT. Second, the exhaust pressure dynamics is expressed as follows:

$$\frac{dP_{exh}}{dt} = \frac{RT_{exh}}{V_{exh}}\left(\left(1+\frac{1}{\lambda_s}\right)\eta_{vol}\frac{P_{man}V_{cyl}}{RT_{man}}\frac{N_e}{30} - D_{turb} - \Phi_{wg}\left(\Pi^*_{wg}\right)\frac{P_{exh}}{\sqrt{RT_{exh}}}u_{wg}\right) \tag{22}$$

where $\Phi\left(\Pi^*_{wg}\right) = \sqrt{\frac{2\gamma}{\gamma-1}\left(\Pi^{*\,\frac{2}{\gamma}}_{wg} - \Pi^{*\,\frac{\gamma+1}{\gamma}}_{wg}\right)}$ with $\Pi^*_{wg} \triangleq \max\left(\frac{P_{dt}}{P_{exh}}, \left(\frac{2}{\gamma+1}\right)^{\frac{\gamma}{\gamma-1}}\right)$ and the gas flow through the turbine $D_{turb} \triangleq \mathrm{LUT}_{D_{turb}}\left(N_{tc}, \frac{P_{dt}}{P_{exh}}\right)$ is given by LUT. Another turbine gas flow model based on the standard equation for compressible flow across an orifice is also available in [33]. Third, the dynamics of the turbocharger can be modeled as

$$\frac{d}{dt}\left(\frac{1}{2}J_{tc}N_{tc}^2\right) = \mathbb{P}_{turb} - \mathbb{P}_{comp} \tag{23}$$

where the powers of the turbine and the compressor are given by

$$\begin{cases} \mathbb{P}_{turb} = D_{turb}C_pT_{exh}\eta_{turb}\left(1 - \Pi_{turb}^{\frac{1-\gamma}{\gamma}}\right) \\ \mathbb{P}_{comp} = D_{comp}C_pT_{amb}\dfrac{1}{\eta_{comp}}\left(\Pi_{comp}^{\frac{\gamma-1}{\gamma}} - 1\right) \end{cases} \tag{24}$$

In expression (24), the following quantities are given by LUTs $\eta_{turb} \triangleq \mathrm{LUT}_{\eta_{turb}}\left(N_{tc}, \frac{P_{dt}}{P_{exh}}\right)$, $\Pi_{comp} \triangleq \mathrm{LUT}_{\Pi_{comp}}(N_{tc}, D_{comp})$, and $\eta_{comp} \triangleq \mathrm{LUT}_{\eta_{comp}}\left(N_{tc}, \frac{P_{boost}}{P_{amb}}\right)$. From (23) and (24), the turbocharger dynamics can be rewritten as follows:

$$\frac{d}{dt}\left(\frac{1}{2}J_{tc}N_{tc}^2\right) = D_{turb}C_pT_{exh}\eta_{turb}\left(1 - \Pi_{turb}^{\frac{1-\gamma}{\gamma}}\right) - D_{comp}C_pT_{amb}\frac{1}{\eta_{comp}}\left(\Pi_{comp}^{\frac{\gamma-1}{\gamma}} - 1\right) \tag{25}$$

From (21), (22), and (25), the dynamical model of turbocharged air system is given as

$$\begin{cases} \dfrac{dP_{man}}{dt} = -\eta_{vol}\dfrac{V_{cyl}}{V_{man}}\dfrac{N_e}{30}P_{man} + \dfrac{P_{boost}\sqrt{RT_{man}}}{V_{man}}\Phi_{thr}(\Pi_{thr}^*)u_{thr} \\ \dfrac{dP_{exh}}{dt} = \dfrac{RT_{exh}}{V_{exh}}\left(\left(1 + \dfrac{1}{\lambda_s}\right)\eta_{vol}\dfrac{P_{man}V_{cyl}}{RT_{man}}\dfrac{N_e}{30} - D_{turb} - \Phi_{wg}\left(\Pi_{wg}^*\right)\dfrac{P_{exh}}{\sqrt{RT_{exh}}}u_{wg}\right) \\ \dfrac{d}{dt}\left(\dfrac{1}{2}J_{tc}N_{tc}^2\right) = D_{turb}C_pT_{exh}\eta_{turb}\left(1 - \Pi_{turb}^{\frac{1-\gamma}{\gamma}}\right) - D_{comp}C_pT_{amb}\dfrac{1}{\eta_{comp}}\left(\Pi_{comp}^{\frac{\gamma-1}{\gamma}} - 1\right) \end{cases} \tag{26}$$

The following features of the turbocharged engine model, directly related to the proposed control solution, should be highlighted [32].

1. This system is highly nonlinear and apparently complex for control design.

2. There are two control inputs (throttle and wastegate) and only one output of interest, the intake pressure, which is directly related to the engine torque.

3. The relation between the wastegate and the intake pressure is not direct.

4. Note that the most commonly available sensors on series production vehicles are found in the intake side of the engine, i.e., the pressure and temperature in the upstream of the compressor (P_{amb}, T_{amb}), the boost pressure P_{boost}, the mass airflow through the compressor D_{comp}, the intake pressure and temperature (P_{man}, T_{man}), and the engine speed N_e.

4.2 MIMO control design

Most of the existing controllers in the open literature, not only aforementioned available measures of engine intake side but also several other signals coming from the exhaust side, i.e., P_{exh}, T_{exh}, P_{dt}, and N_{tc}, are needed for control implementation.

However, these signals are not measured in commercial vehicles and usually assumed to be estimated by estimators/observers. To avoid this practical issue, here these variables are approximated by their static LUTs issued from the data measured in steady-state conditions in the test bench. Hence, we can reduce the number of costly vehicle sensors or/and complex observers. Concretely, the following LUTs are constructed:

$$\begin{cases} P_{exh} = \mathrm{LUT}_{P_{exh}}(N_e, P_{man}) \\ T_{exh} = \mathrm{LUT}_{T_{exh}}(N_e, D_{cyl}) \\ P_{dt} = \mathrm{LUT}_{P_{dt}}(N_e, D_{cyl}) \\ N_{tc} = \mathrm{LUT}_{N_{tc}}(\Pi_{comp}, D_{comp}) \end{cases} \tag{27}$$

Remark from (27) that all the inputs of respective LUTs P_{exh}, T_{exh}, P_{dt}, N_{tc} can be obtained with available vehicle sensors. The approximations in (27) are reasonable since SI engines operate at stoichiometric conditions, which implies that all exhaust variables are highly correlated to the in-cylinder air mass flow (or intake pressure). Note also that although such an approximation may introduce some estimation errors, especially during the transient phases, the proposed robust control approach is expected to compensate these errors.

We now focus on the robust control design. Apart from the output of interest $y_{man} = P_{man}$, we virtually introduce the second output $y_{exh} = P_{exh}$ to facilitate the control design task. Note that the goal is only to track the intake pressure reference $P_{man,ref}$. Moreover, we do not have the exhaust pressure reference $P_{exh,ref}$ in practice. However, by means of LUT in (27), we can impose that $P_{exh,ref} = \mathrm{LUT}_{P_{exh}}(N_e, P_{man,ref})$ and then if P_{exh} converges to $P_{exh,ref}$, it implicitly makes P_{man} converge to $P_{man,ref}$. Hence, both outputs P_{man} and P_{exh} are used to track the intake pressure reference. For engine control design, we consider the two pressure dynamics in (21) and (22), which can be rewritten in the following compact form:

$$\begin{cases} \dot{P}_{man} = K_{man}(D_{thr} - D_{cyl}) \triangleq f_{thr} + g_{thr} u_{thr} \\ \dot{P}_{exh} = K_{exh}(K_{fuel} D_{cyl} - D_{turb} - D_{wg}) \triangleq f_{wg} + g_{wg} u_{wg} \\ y_{man} \triangleq P_{man}, \quad y_{exh} \triangleq P_{exh} \end{cases} \tag{28}$$

where

$$\begin{cases} K_{man} = \dfrac{RT_{man}}{V_{man}}, \quad K_{exh} = \dfrac{RT_{exh}}{V_{exh}}, \quad K_{fuel} = \left(1 + \dfrac{1}{\lambda_s}\right) \\ D_{cyl} = K_{cyl} P_{man}, \quad K_{cyl} = \eta_{vol} \dfrac{V_{cyl}}{V_{man}} \dfrac{N_e}{30} \\ f_{thr} = -K_{man} D_{cyl}, \quad f_{wg} = K_{exh}(K_{fuel} D_{cyl} - D_{turb}) \\ g_{thr} = K_{man} \dfrac{P_{boost}}{\sqrt{RT_{man}}} \Phi_{thr}(\Pi^*_{thr}), \quad g_{wg} = -K_{exh} \dfrac{P_{exh}}{\sqrt{RT_{exh}}} \Phi_{wg}(\Pi^*_{wg}) \end{cases} \tag{29}$$

Now, the feedback linearization technique is applied to control the nonlinear system (28). To this end, let us compute the time derivatives of the outputs as

$$\begin{cases} \dot{y}_{man} = \dot{P}_{man} = f_{thr} + g_{thr} u_{thr} = v_{man} \\ \dot{y}_{exh} = \dot{P}_{exh} = f_{wg} + g_{wg} u_{wg} = v_{exh} \end{cases} \tag{30}$$

The two control inputs u_{thr}, u_{wg} appear respectively in $\dot{y}_{man}$, $\dot{y}_{exh}$; the signals v_{man} and v_{exh} are two new manipulated inputs. Using an integral structure for tracking control purposes, the following linearized system is derived from (28):

$$\begin{cases} \dot{y}_{man} = v_{man} \\ \dot{y}_{exh} = v_{exh} \\ \dot{x}_{\text{int}} = y_{man,ref} - y_{man} \end{cases} \tag{31}$$

with the feedback linearization control laws

$$\begin{cases} u_{thr} = -\dfrac{f_{thr}}{g_{thr}} + \dfrac{1}{g_{thr}} v_{man} \\ u_{wg} = -\dfrac{f_{wg}}{g_{wg}} + \dfrac{1}{g_{wg}} v_{exh} \end{cases} \tag{32}$$

Define $x \triangleq [y_{man}, y_{exh}, x_{\text{int}}]^T$, $v \triangleq [v_{man}, v_{exh}]^T$, and suppose that system (28) is subject to modeling errors $\Delta(x)$ caused by nonlinearities $f_{thr}, g_{thr}, f_{wg}, g_{wg}$ and the approximation by using LUTs. Then, the linearized system (31) is rewritten as

$$\dot{x} = \begin{pmatrix} 0 & 0 & 0 \\ 0 & 0 & 0 \\ -1 & 0 & 0 \end{pmatrix} x + \begin{pmatrix} 1 & 0 \\ 0 & 1 \\ 0 & 0 \end{pmatrix} v + \begin{pmatrix} 0 \\ 0 \\ 1 \end{pmatrix} y_{man,ref} + \Delta(x) \tag{33}$$

We assume that $\Delta(x) \leq \delta^2 x^T H^T H x$. Theorem 1 can be applied to design v_{man} and v_{exh}. Selecting $H = I$, $\lambda_1 = 1$, $\lambda_2 = \lambda_3 = 0$, and $\overline{\delta} = 0.9$, then we obtain the following control law:

$$v = -Kx = -\begin{bmatrix} 110.3 & 0 & -4052 \\ 0 & 48.9 & 0 \end{bmatrix} x \tag{34}$$

and $\delta = 0.9983$, which is larger than prescribed value of $\overline{\delta}$, as expected.

The stability analysis of the internal dynamics is necessary to make sure that the state N_{tc}^2 is well behaved. To this end, the turbocharger dynamics (25) is rewritten in the form

$$\frac{d}{dt}\left(N_{tc}^2\right) = K_{turb} D_{turb} - K_{comp} D_{comp} \tag{35}$$

where

$$K_{turb} \triangleq \frac{2}{J_{tc}} C_p T_{exh} \eta_{turb} \left(1 - \Pi_{turb}^{\frac{1-\gamma}{\gamma}}\right); \quad K_{comp} \triangleq \frac{2}{J_{tc}} C_p T_{amb} \frac{1}{\eta_{comp}} \left(\Pi_{comp}^{\frac{\gamma-1}{\gamma}} - 1\right) \tag{36}$$

Moreover, we obtain from (28) and (30) that

$$D_{turb} = K_{fuel} D_{cyl} - D_{wg} - \frac{v_{exh}}{K_{exh}} \tag{37}$$

It follows from (35) and (37) that

$$\frac{d}{dt}\left(N_{tc}^2\right) = -K_{turb} D_{wg} - K_{comp} D_{comp} + K_{turb} K_{fuel} D_{cyl} - \frac{K_{turb}}{K_{exh}} v_{exh} \tag{38}$$

Note that $P \triangleq [P_{man}, P_{exh}]^T$ can be considered as the input vector of system (38). Then, it follows that

$$\frac{d}{dt}(N_{tc}^2) < \left(K_{turb}K_{fuel}K_{cyl}P_{man} + \frac{K_{turb}}{K_{exh}}K_{(2,2)}P_{exh}\right) - K_{comp}D_{comp}$$

$$\leq \sqrt{(K_{turb}K_{fuel}K_{cyl})^2 + \left(\frac{K_{turb}}{K_{exh}}K_{(2,2)}\right)^2}\|P\| - K_{comp}D_{comp} \triangleq \alpha(t)\|P\| - \beta(N_{tc}^2) \tag{39}$$

Since α is bounded and the function $\beta(\cdot)$ is of class $\mathcal{K}^\infty$ (see **Figure 2**). Hence, we can conclude that system (35) is input-to-state stable [34].

Hereafter, the controller designed in this subsection is called *conventional MIMO controller*.

4.3 Fuel-optimal control strategy

We have designed in this work *conventional MIMO controller* with two inputs, throttle and wastegate, and two outputs: intake pressure and exhaust pressure for the whole engine operating zone. From the viewpoint of energy efficiency, this controller is not *optimal* in the sense of energy losses minimization. Indeed, the wastegate should be opened as much as possible at a given operating point to minimize the pumping losses [15]. This leads to the control strategy proposed in [16], i.e., in low-load zone, only the throttle is used to track the intake pressure and the wastegate is widely open, and in high-load zone, the wastegate is solely activated to control the pressure and the throttle is widely open in this case. To fully take into account the above fuel-optimal strategy, we propose the so-called fuel-optimal controller for turbocharged air system of a SI engine. This novel controller is directly derived from *conventional MIMO controller*, and they both have the same control law (34). The idea is presented in the sequel.

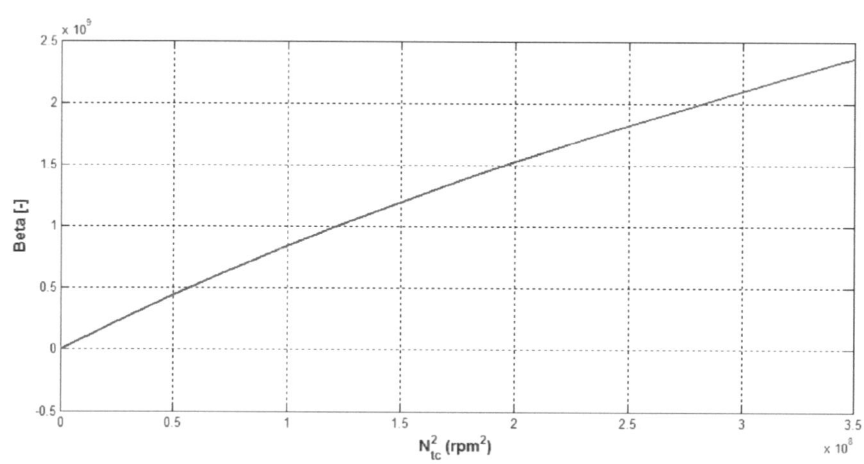

Figure 2.
Behavior of $\beta(N_{tc}^2)$ function.

Let us recall the engine model (28) as

$$\begin{cases} \dot{P}_{man} = K_{man}\left(D_{thr} - D_{cyl}\right) \triangleq f_{thr} + g_{thr}u_{thr} = v_{man} \\ \dot{P}_{exh} = K_{exh}\left(K_{fuel}D_{cyl} - D_{turb} - D_{wg}\right) \triangleq f_{wg} + g_{wg}u_{wg} = v_{exh} \\ y_{man} \triangleq P_{man} \\ y_{exh} \triangleq P_{exh} \end{cases} \tag{40}$$

It follows from the second equation of (40) that

$$D_{cyl} = \frac{v_{exh}}{K_{exh}K_{fuel}} + \frac{D_{turb}}{K_{fuel}} + \frac{D_{wg}}{K_{fuel}} \tag{41}$$

Then, the intake pressure dynamics can be also rewritten as

$$\dot{P}_{man} = K_{man}\left(D_{thr} - \frac{v_{exh}}{K_{exh}K_{fuel}} - \frac{D_{turb}}{K_{fuel}} - \frac{D_{wg}}{K_{fuel}}\right) \tag{42}$$

or equivalently

$$\dot{P}_{man} = g_{thr}u_{thr} - \frac{K_{man}}{K_{exh}K_{fuel}}v_{exh} - \frac{K_{man}}{K_{fuel}}D_{turb} + \frac{K_{man}}{K_{exh}K_{fuel}}g_{wg}u_{wg} = v_{man} \tag{43}$$

The novel *fuel-optimal controller* is directly derived from the above expression. To this end, the whole engine operating range is divided into three zones according to two predefined intake pressure thresholds P_{man1} and P_{man2}.

1. *Zone 1* (low-load zone $P_{man} \le P_{man1}$): The wastegate is widely open, and the throttle is solely used to track the intake pressure reference. Let $S_{wg,\max}$ be the maximal opening section of the wastegate. The implemented actuator control laws are in this case

$$\begin{cases} u_{thr} = \dfrac{1}{g_{thr}}\left(\dfrac{K_{man}}{K_{exh}K_{fuel}}v_{exh} + \dfrac{K_{man}}{K_{fuel}}D_{turb} - \dfrac{K_{man}}{K_{exh}K_{fuel}}g_{wg}S_{wg,\max} + v_{man}\right) \\ u_{wg} = S_{wg,\max} \end{cases} \tag{44}$$

2. *Zone 2* (middle-load zone $P_{man1} < P_{man} < P_{man2}$): Both throttle and wastegate are simultaneously used to control the intake pressure. In this case, the implemented actuator control laws are exactly the feedback linearization laws in (32), which are recalled here

$$\begin{cases} u_{thr} = -\dfrac{f_{thr}}{g_{thr}} + \dfrac{1}{g_{thr}}v_{man} \\ u_{wg} = -\dfrac{f_{wg}}{g_{wg}} + \dfrac{1}{g_{wg}}v_{exh} \end{cases} \tag{45}$$

3. *Zone 3* (high-load zone $P_{man} \ge P_{man2}$): The throttle is fully opened, and only the wastegate is activated to control the intake pressure which is approximated by the boost pressure P_{boost}. The implemented actuator control laws are

$$\begin{cases} u_{thr} = S_{thr,\max} \\ u_{wg} = \dfrac{K_{exh}K_{fuel}}{K_{man}g_{thr}}\left(\dfrac{K_{man}}{K_{exh}K_{fuel}}v_{exh} + \dfrac{K_{man}}{K_{fuel}}D_{turb} - g_{thr}S_{thr,\max} + v_{man}\right) \end{cases} \tag{46}$$

where $S_{thr,\max}$ is the maximal opening section of the throttle.

Several remarks can be reported for this actuator scheduling strategy. First, since the input vector $v \triangleq [v_{man}, v_{exh}]^T$ is the same for all three zones, then the dynamics of N_{tc}^2 defined in (35) is always input-to-state stable with this strategy since it does not directly depend on the real control inputs u_{thr} and u_{wg} of the turbocharged air system. Second, the exhaust pressure dynamics can be rewritten as

$$\begin{aligned} \dot{P}_{exh} &= K_{exh}K_{fuel}K_{cyl}P_{man} - K_{exh}\left(\frac{P_{exh}}{\sqrt{RT_{exh}}}\Phi_{wg}\left(\Pi_{wg}^{*}\right)u_{wg} + D_{turb}\right) \\ &< K_{exh}K_{fuel}K_{cyl}P_{man} - K_{exh}D_{turb} \triangleq \theta(P_{man}) - \tau(P_{exh}) \end{aligned} \tag{47}$$

Note that the functions $\theta(\cdot)$ and $\tau(\cdot)$ are of class $\mathcal{K}^\infty$ and then the exhaust pressure dynamics is always input-to-state stable with respect to P_{man}. Third, it follows from the above remarks that if the intake pressure tracking performance is guaranteed, then all other variables of the turbocharged air system (26) will be well behaved within three operating zones. Fourth, the model-based *fuel-optimal controller* is based on a dummy switching strategy because no switching model has been used in this approach. Fifth, the pressure thresholds P_{man1} and P_{man2} separating the three zones are *freely* chosen thanks to the propriety of the above third remark. However, the values of P_{man1}, P_{man2} are usually chosen very close for engine efficiency benefits.

Note that *fuel-optimal controller* is different from other existing approaches in the literature. As the approach proposed in [16], this novel controller is a MIMO nonlinear controller which can guarantee the closed-loop stability of the whole turbocharged air system. However, the novel *fuel-optimal controller* is much simpler, and the middle-load zone (Zone 2) is very easily introduced to improve the torque response at high load while maintaining the maximum possible advantage of *fuel-optimal* concept in [15]. The scheduling strategy of *fuel-optimal controller* has also appeared in [35]. However, the control approach in [35] is based on a decentralized linear scheduling PI controller. In addition, the throttle is only *passively* activated in Zone 2, that is, the throttle control is maintained at a constant value obtained from calibration for each operating point of the engine. Moreover, the authors did not show how to choose the intake pressure thresholds and in particular how this choice will effect on the control design. Compared with the control approach in [36] which is also based on feedback linearization, our controller does not need any model simplification task, e.g., neglecting pressure dynamics with respect to turbocharger dynamics according to singular perturbation theory and approximating the turbocharger square speed as a linear function of intake pressure. Note also that the same simplification procedure is carried out for the approach in [16, 17, 20]. Moreover, in [36], the wastegate and the throttle are separately controlled, and the approach cannot take into account the mid-load zone.

4.4 Simulation results and analysis

Hereafter, a series of trials are performed on an engine simulator designed under commercial AMESim platform [18] to show the effectiveness of the proposed method for both cases: *conventional MIMO controller* and *fuel-optimal controller*.

For the sake of clarity, the two commands (throttle, wastegate) are normalized. Then, the control input constraints become $0 \leq \overline{u}_{thr}, \overline{u}_{wg} \leq 100\%$. When $\overline{u}_{thr} = 100\%$ (respectively, $\overline{u}_{wg} = 0\%$), it means that the throttle (resp. wastegate) is fully open. On the reverse, when $\overline{u}_{thr} = 0\%$ (respectively, $\overline{u}_{wg} = 100\%$), the throttle (resp. wastegate) is fully closed. Before starting, note that the proposed controller is easily tuned with only one parameter, the desired robustness bound $\overline{\delta}$ which is the same for all following simulations. The pressure thresholds are chosen as $P_{man,1} = 0.9$ bar and $P_{man,2} = 1.2$ bar.

4.4.1 Comparison between conventional MIMO *control and fuel-optimal control*

Figures 3 and **4** represent the intake pressure tracking performance and the corresponding actuator commands for *conventional MIMO controller* and *fuel-optimal controller*, respectively. *Conventional MIMO controller* simultaneously uses both actuators to track the intake pressure, while these actuators are optimally scheduled by the strategy described in SubSection 4.3 with *fuel-optimal controller*. The wastegate is opened very little with *conventional MIMO controller* so that the boost potential of the turbocharger can be fully exploited. Hence, the closed-loop time response with this controller is faster than the one of *fuel-optimal controller* in middle- and high-load zones. Moreover, although *conventional MIMO controller* can be used to improve the torque response (drivability), this controller is not optimal in terms of fuel consumption compared with *fuel-optimal controller* as pointed out in **Figure** 5. The pumping losses with *fuel-optimal controller* are almost lower than the ones with *conventional MIMO controller* at every time. Observe that the pumping losses with *fuel-optimal controller* are very low at high intake pressure.

Since the goal of this work is to design a controller minimizing the energy losses, only results with the *fuel-optimal controller* will be presented in the rest of this chapter.

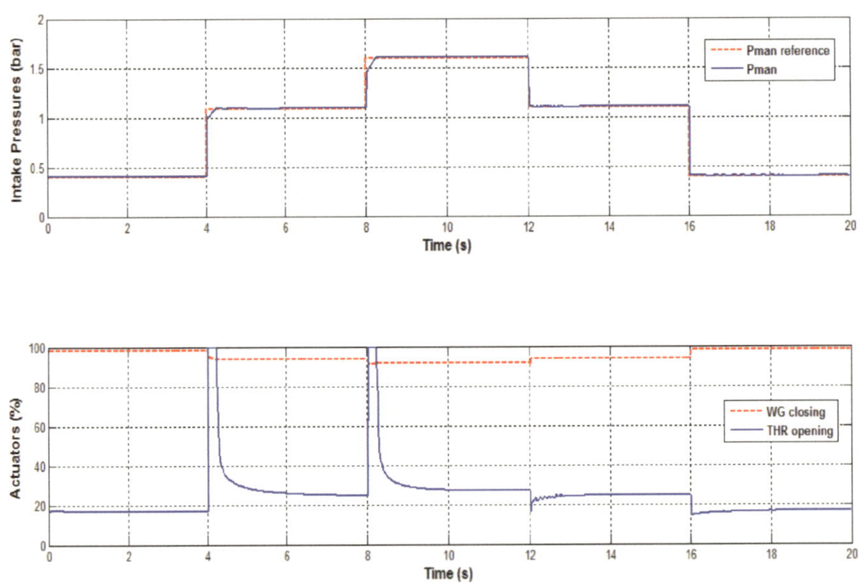

Figure 3.
Pressure tracking performance (up) and corresponding actuator commands (bottom) with conventional MIMO controller *at* $N_e = 2000$ rpm.

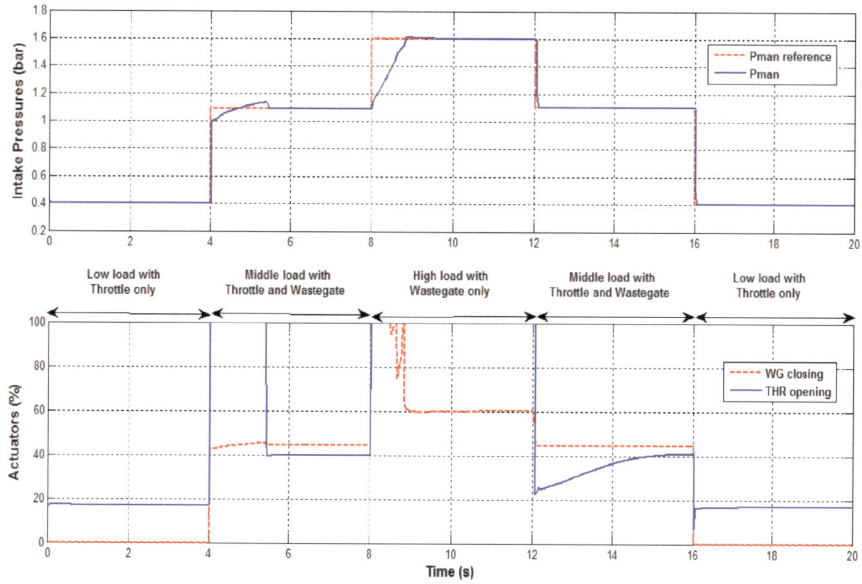

Figure 4.
Pressure tracking performance (up) and corresponding actuator commands (bottom) with fuel-optimal controller *at* $N_e = 2000$ rpm.

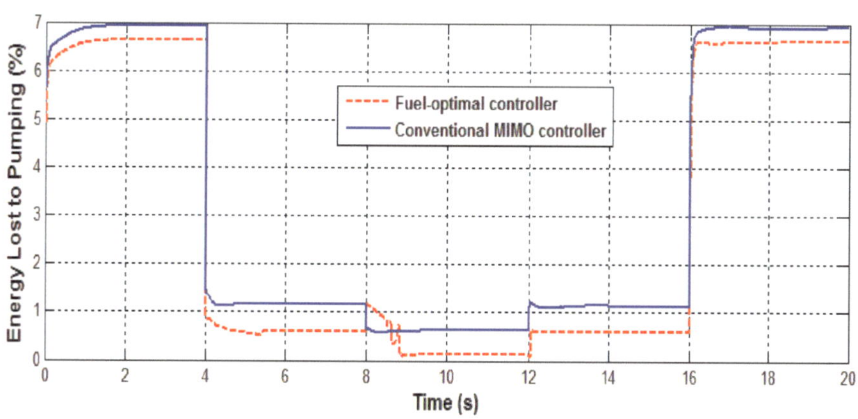

Figure 5.
Comparison of engine pumping losses between conventional MIMO controller *and* fuel-optimal controller *at* $N_e = 2000$ rpm.

4.4.2 Fuel-optimal controller performance at different engine speeds

The trajectory tracking of the intake pressure at different engine speeds is shown in **Figure 6**. The following comments need to be made regarding these results. First, the tracking performance is very satisfying over the whole operating range. The wastegate command is very aggressive during the turbocharger transients; it hits the constraints and then stabilizes to track the boost pressure. This fact allows compensating the slow dynamics of the turbocharger. Moreover, this behavior can be easily tuned with the parameter $\overline{\delta}$, i.e., a smaller value of $\overline{\delta}$ leads to the faster time response; however the robustness bound will be reduced. Second, the controller does not generate any overshoot in the considered operating range which is also a very important property for the driving comfort.

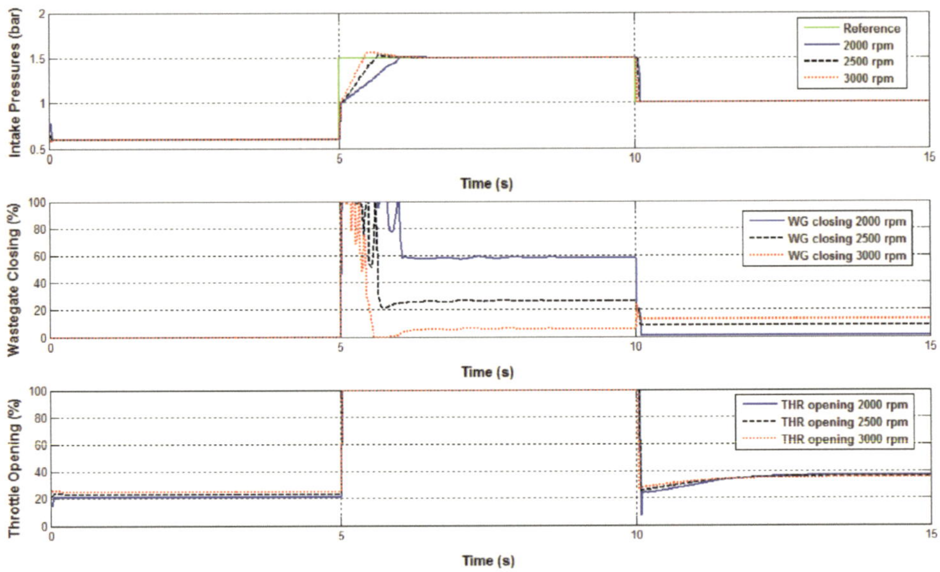

Figure 6.
Intake pressure tracking performance (up) with corresponding wastegate commands (middle) and throttle commands (bottom) at different engine speeds.

4.4.3 Vehicle transients

The closed-loop responses during the vehicle transient are presented in **Figure 7**. It can be noticed that the *fuel-optimal controller* is perfectly able to guarantee a very good tracking performance even with the important variation of the engine speed (which represents the vehicle transient).

All of the above test scenarios and the corresponding results confirm the effectiveness of the proposed approach over the whole engine operating range. It is

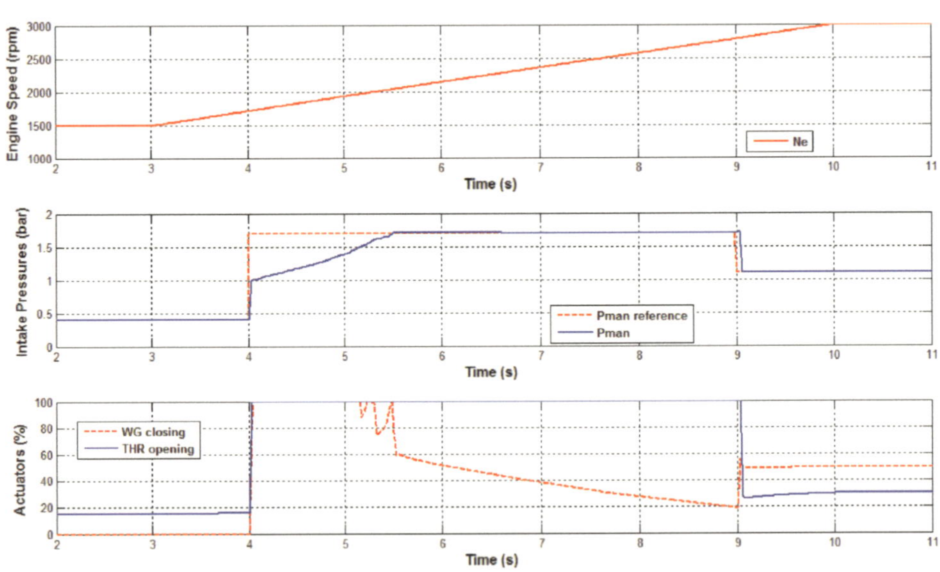

Figure 7.
Variation of engine speed (up) and pressure tracking performance (middle) with corresponding actuator commands (bottom) for a vehicle transient.

emphasized again that the same controller gain is used for both controllers in all simulations. Therefore, the proposed approach requires very limited calibration effort.

5. Concluding remarks

A new robust control design has been proposed to handle the modeling uncertainty and/or disturbances, known as one of major drawbacks of feedback linearization. Compared to the existing results, the proposed method provides a simple and constructive design procedure which can be recast as an LMI optimization problem. Hence, the controller feedback gain is effectively computed.

In terms of application, an original idea has been proposed to control the turbocharged air system of a SI engine. Several advantages of this approach can be summarized as follows. First, the second virtual output $y_{exh} \triangleq P_{exh}$ is introduced by means of LUT, and this fact drastically simplifies the control design task. Second, the resulting nonlinear control law is easily implementable. Third, offline engine data of the test bench is effectively reused and exploited for engine control development so that the number of sensors and/or observers/estimators could be significantly reduced. Finally, the controller is robust with respect to modeling uncertainties/disturbances, and its feedback gain can be effectively computed through a convex optimization problem with available numerical solvers. Despite its simplicity, the proposed controller can provide very promising results for both control strategies of turbocharged air system, i.e., to improve the drivability with *conventional MIMO controller* or to optimize the fuel consumption with *fuel-optimal controller*. Future works focus on the real-time validation of the proposed fuel-optimal control strategy with an engine test bench.

Author details

Anh-Tu Nguyen*, Thierry-Marie Guerra and Jimmy Lauber
Laboratory LAMIH UMR CNRS 8201, Université Polytechnique Hauts-de-France, Valenciennes, France

*Address all correspondence to: nguyen.trananhtu@gmail.com

References

[1] Moulin P. Air systems modeling and control for turbocharged engines [PhD thesis]. MINES Paristech; 2010

[2] Guzzella L, Onder C. Introduction to Modeling and Control of Internal Combustion Engine Systems. Berlin Heidelberg: Springer-Verlag; 2004

[3] Daubler L, Bessai C, Predelli O. Tuning strategies for online-adaptive PI controller. Oil and Gas Science and Technology. 2007;**62**(4):493-500

[4] Karnik A, Buckland J, Freudenberg J. Electronic throttle and wastegate control for turbocharged gasoline engines. American Control Conference. Portland; 2005. pp. 4434-4439

[5] Jung M, Glover K, Christen U. Comparison of uncertainty parameterisations for H-infinity robust control of turbocharged diesel engines. Control Engineering Practice. 2005; **13**(1):15-25

[6] Wei X, del Re L. Gain scheduled H∞ control for air path systems of diesel engines using LPV techniques. IEEE Transactions on Control Systems Technology. 2007;**15**(3):406-415

[7] Utkin V, Chang H, Kolmanovsky I, Cook J. Sliding mode control for variable geometry turbocharged diesel engines. In: American Control Conference. Chicago; 2000. pp. 584-588

[8] Ortner P, del Re L. Predictive control of a diesel engine air path. IEEE Transactions on Control Systems Technology. 2007;**15**(3):449-456

[9] Kim S, Jin H, Choi S. Exhaust pressure estimation for diesel engines equipped with dual-loop EGR and VGT. IEEE Transactions on Control Systems Technology. 2017;**26**(2):382-392

[10] Wang H, Bosche J, Tian Y, El Hajjaji A. Two loop based dynamical feedback stabilization control of a diesel engine with EGR & VGT. In: 50th IEEE Conference on Decision and Control and European Control Conference (CDC-ECC). Orlando, USA; 2011. pp. 1596-1601

[11] Zhang Y, Lu G, Xu H, Li Z. Tuneable model predictive control of a turbocharged diesel engine with dual loop exhaust gas recirculation. Proceedings of the Institution of Mechanical Engineers, Part D. 2018; **232**(8):1105-1120

[12] Larguech S, Aloui S, El Hajjaji A, Chaari A. Fuzzy sliding mode control for turbocharged diesel engine. Journal of Dynamic Systems, Measurement, and Control. 2016;**138**(1):011009

[13] Gong X, Wang Y, Chen H, Hu Y. Double Closed-Loop Controller Design for Boost Pressure Control of Turbocharged Gasoline Engines. Vol. 7. IEEE Access; 2019. pp. 97333-97342

[14] Hu Y, Chen H, Wang P, Chen H, Ren L. Nonlinear model predictive controller design based on learning model for turbocharged gasoline engine of passenger vehicle. Mechanical Systems and Signal Processing. 2018; **109**:74-88

[15] Eriksson L, Frei S, Onder C, Guzzella L. Control and optimization of turbocharged SI engines. 15th IFAC World Congress. Barcelona, Spain; 2002

[16] Nguyen A-T, Dambrine M, Lauber J. Lyapunov-based robust control design for a Class of switching nonlinear systems subject to input saturation: Application to engine control. IET Control Theory and Applications. 2014a;**8**(17):1789-1802

[17] Nguyen A-T, Lauber J, Dambrine M. Robust H∞ control design for switching uncertain system: Application for

turbocharged gasoline air system control. In: 51st Annual Conference on Decision and Control (CDC). Maui, Hawaii, USA: IEEE; 2012a. pp. 4265-4270

[18] Nguyen A-T, Lauber J, Dambrine M. Optimal control based algorithms for energy management of automotive power systems with battery/supercapacitor storage devices. Energy Conversion and Management. 2014b;**87**:410-420

[19] Nguyen A-T, Lauber J, Dambrine M. Multiobjective control design for turbocharged spark ignited air system: A switching Takagi-Sugeno model approach. In: American Control Conference (ACC). Washington, DC, USA: IEEE; 2013. pp. 2866-2871

[20] Nguyen A-T, Lauber J, Dambrine M. Switching fuzzy control of the air system of a turbocharged gasoline engine. In: IEEE International Conference on Fuzzy Systems (FUZZ-IEEE). Brisbane: IEEE; 2012b. pp. 1-7

[21] Nguyen A-T, Sugeno M, Campos V, Dambrine M. LMI-based stability analysis for piecewise multi-affine systems. IEEE Transactions on Fuzzy Systems. 2016;**25**(3):707-714

[22] Nguyen A-T, Taniguchi T, Eciolaza L, Campos V, Palhares R, Sugeno M. Fuzzy control systems: Past, present and future. IEEE Computational Intelligence Magazine. 2019;**14**(1):56-68

[23] Isidori A. Nonlinear Control Systems. London: Springer Verlag; 1989

[24] Khalil H. Nonlinear Systems. 3rd ed. Upper Saddle River, NJ: Prentice Hall; 2002

[25] Sastry S. Nonlinear Systems. New York: Springer-Verlag; 1999

[26] Ha I, Gilbert E. Robust tracking in nonlinear systems. IEEE Transactions on Automatic Control. 1987;**32**(9): 763-771

[27] Kravaris C, Palanki S. A Lyapunov approach for robust nonlinear state feedback synthesis. IEEE Transactions on Automatic Control. 1988;**33**(12):1188-1191

[28] Boyd S, Ghaoui L, Feron E, Balakrishnan V. Linear Matrix Inequalities in System and Control Theory. Philadelphia: Society for Industrial and Applied Mathematics (SIAM); 1994

[29] Šiljak D, Stipanović D. Robust stabilization of nonlinear systems: The LMI approach. Mathematical Problems in Engineering. 2000;**6**(5):461-493

[30] Gahinet P, Nemirovski A, Laub A, Chilali M. LMI Control Toolbox. Apple Hill Drive: The Math Works Inc.; 1995

[31] Lofberg J. YALMIP: A toolbox for modeling and optimization in MATLAB. In: IEEE International Symposium on Computer Aided Control Systems Design. Taipei; 2004. pp. 284-289

[32] Nguyen A-T. Advanced Control Design Tools for Automotive Applications [PhD thesis]. University of Valenciennes et du Hainaut-Cambresis; 2013

[33] Eriksson L. Modeling and control of turbocharged SI and DI engines. Oil and Gas Science and Technology. 2007; **62**(4):523-538

[34] Sontag E, Wang Y. On characterizations of the input to state stability property. Systems and Control Letters. 1995:351-359

[35] Gorzelic P, Hellström E, Stefanopoulou A, Jiang L, Gopinath S. A coordinated approach for throttle and wastegate control in turbocharged spark ignition engines. 24th Chinese Control and Decision Conference. Taiyuan, China; 2012. pp. 1524-1529

[36] Moulin P, Chauvin J. Modeling and control of the air system of a turbocharged gasoline engine. Control Engineering Practice. 2011;**19**(3):287-297

Chapter 6

Real-Time Simulation of Efficient Energy Management Algorithms for Electric Vehicle Chargers

Santhosh Thuttampatty Krishnamoorthy, Suthanthira Vanitha Narayanan and Ramkumar Kannan

Abstract

Transportation electrification is happening at a rapid pace around the globe in response to the climate change mitigation measures taken by the regulatory agencies to curb tailpipe emissions. As the electric vehicle technology evolved, the size of on-board storage units has increased, which require charging from an external energy source. Renewable charging of electric vehicles is an attractive option to reduce the carbon footprint of an electric vehicle. The intermittent nature of the renewables necessitates a storage unit to provide continuous power. With a battery complementing solar generation, a power converter is deployed to interface these sources and storage units with the electric vehicle for charging. The converter shall now have to operate to quench the charging requirements by sourcing power from solar generation and storage elements. The converter also has to capture the generated solar power during the non-charging period and store it in the battery. All these functional requirements demand a robust energy management strategy to utilize all available sources and storage units efficiently without compromising load requirements. A Stateflow-based energy management algorithm for a three-port converter is proposed in this work. The proposed algorithm is implemented using OPAL-RT, and the real-time simulation results are presented.

Keywords: energy management, electric vehicle, renewable charging, multiport converter, real-time simulation, OPAL-RT

1. Introduction

Transportation accounts for more than 30% of the atmospheric particulate emissions around the globe. The unprecedented increase in environmental pollution leads to climate change which affects the biodiversity and environment of this planet. Sensing the ill effects of global warming, several steps have been taken to minimize the effect of environmental pollutants. Several governments and organizations have imposed stringent emission norms for newly manufactured and old vehicles to curb tailpipe emissions from automobiles [1]. Since fossil fuel-based vehicles cannot be made free from emissions, automotive manufacturers are looking to electrify the transportation section to improve energy efficiency and reduce vehicular emissions.

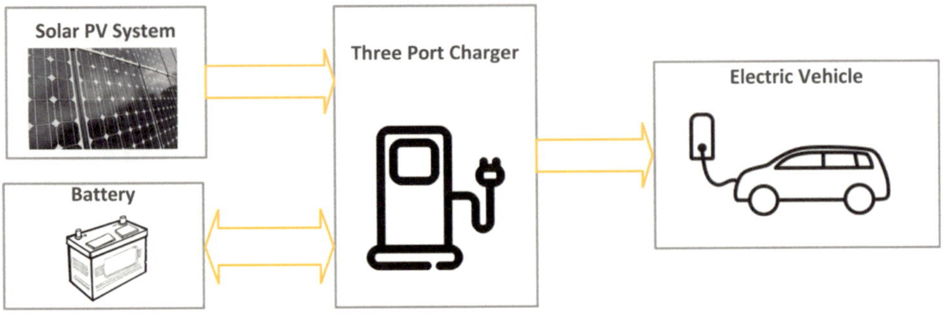

Figure 1.
Block diagram of the three-port charger.

An electric vehicle (EV) is often seen as a viable option to curb atmospheric emission. At the same time, the dependence of charging power sourced from fossil fuels increases the carbon footprint of an electric vehicle. The heated debate is continuing [2–4] in the backdrop of the increase in the global trend in the acceptance of electric vehicle which is reflected in the sales of electric vehicles [5]. As the EV market grows, so is the need for the electric vehicle supply equipment (EVSE). The EVSE, mainly the charger for EVs, is a necessary ancillary growing along with the EV market. The EVSE may be supplying power in AC or DC [6] from different levels. Additionally, there are different charging connectors [7] which need standardization. Most of the charger power is sourced from the grid, which affects the grid stability [8]. The grid connected chargers are predominantly fast chargers [9] which may be inductive [10] or conductive [11]. The dependence of the EV on the grid can be reduced by charging the vehicle from renewable energy sources [12]. The intermittent nature of renewable sources demands the inclusion of storage for improved reliability [13]. The charging station may be a stand-alone charging unit or may be a part of a microgrid [14, 15] which requires an energy management controller [16]. A vast majority of the charger topologies are still grid dependent which undermine the clean energy image of the electric vehicle. This work propose an off-grid stand-alone renewable charger suitable for slow charging. A typical off-grid charger topology is shown in **Figure 1**. The design phase of the charger can be accelerated by deploying real-time simulation [17] which could be used to validate the energy management algorithms in real-time scenarios and generate field-deployable code for rapid prototyping.

2. Electric vehicle chargers and energy management

2.1 Charger topology

A typical charging station with a rooftop solar photovoltaic (PV) generation and a battery considered in this work is shown in **Figure 2**. This topology is modified from an onboard multiport converter proposed in [18] which can be categorized as a level 1 or level 2 charger [19] which supports prolonged charging periods for the vehicular battery. As fast charging is not considered, the grid connection has not been considered. Additionally, the original topology presented in [18] could be used with the front-end rectifier-based grid connected port to support fast charging. The battery-supported solar PV could be operated as an active generator that could be used to power the charging station [20]. The charger topology is non-isolated, suitable for top-up charging of an EV in the workplace and commercial establishments.

2.2 Operating modes

The three-port converter considered in this works has three different operating modes as shown in **Figure 3**. The PV-based active generator has a solar PV unit and a battery. The load port is connected to an EV through an appropriate charging connector. The connector is also used for exchanging the battery system parameters with the charger.

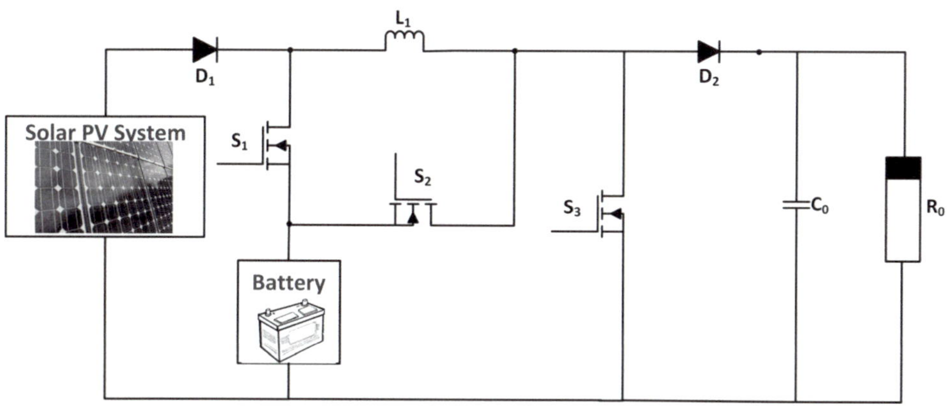

Figure 2.
Circuit diagram of the proposed charger topology.

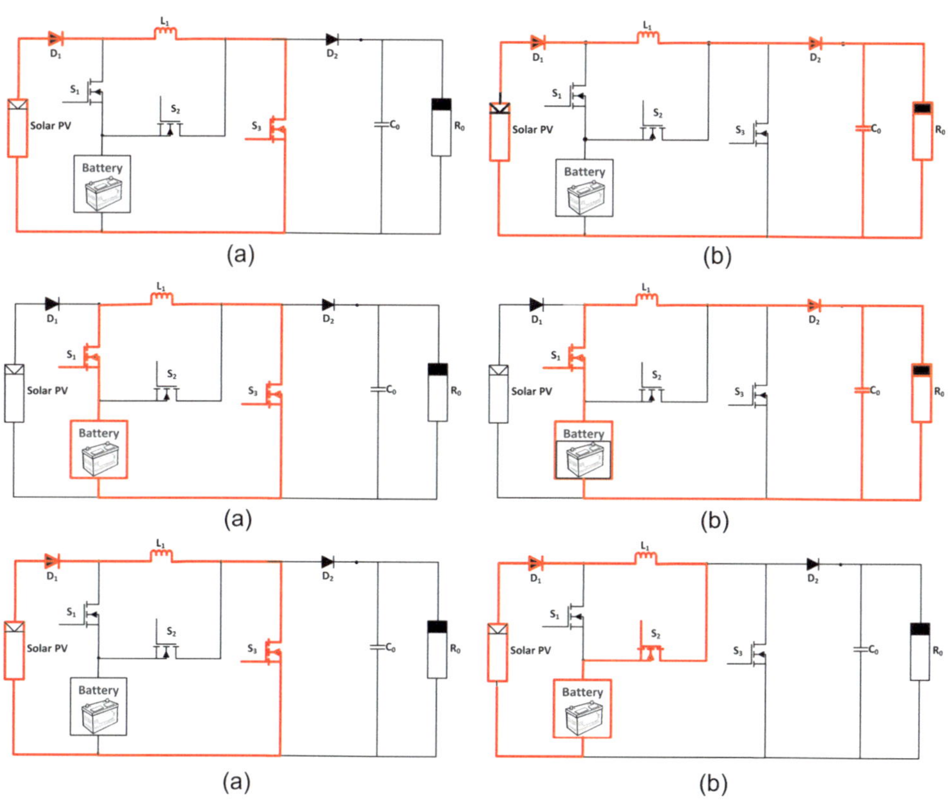

Figure 3.
Different modes and switching states. (a) Mode 1 State 1, (b) Mode 1 State 2; (a) Mode 2 State 1, (b) Mode 2 State 2; (a) Mode 3 State 1, (b) Mode 3 State 2.

2.2.1 Mode 1: solar charging mode

In this mode, the solar power is directly utilized for charging the EV batteries. In the first switching state, the switch S_3 is turned *ON* and the inductor gets charged. The charging current slope is given by

$$\frac{di_{L1}}{dt} = \frac{V_i}{L_1} \tag{1}$$

where V_i is the solar PV voltage and L_1 is the value of the inductor. In the second switching state, the S_3 is turned *off*, and the source voltage combined with the inductor stored voltage is transferred to the load port. The discharging slope for the inductor current is given by

$$\frac{di_{L1}}{dt} = \frac{V_i - V_0}{L_1} \tag{2}$$

where V_0 is the output voltage.

2.2.2 Mode 2: storage charging mode

In this mode, the energy stored in the off-board battery is utilized for charging the EV battery. This mode is suitable when solar PV generation is not sufficient to satisfy the load demand or when the solar power is not available. In the first switching state of Mode 2, the switching devices S_1 and S_3 are turned on simultaneously. The battery voltage (V_b) is used to charge the inductor in the first switching state which is given by

$$\frac{di_{L1}}{dt} = \frac{V_b}{L_1} \tag{3}$$

In the subsequent switching state, the switching device S_3 is turned *off*, while the switching device S_1 is kept *on* continuously. The discharging slope of the inductor current (i_{L1}) is given by

$$\frac{di_{L1}}{dt} = \frac{V_b - V_0}{L_1} \tag{4}$$

2.2.3 Mode 3: surplus storage mode

In this mode, the surplus energy generated by the solar PV is stored in the associated storage batteries, while the charger is idle. These modes serve the dual purpose of energy storage capture during the idle period and support charging when the solar PV generation is not adequate to prove the necessitated charging power. This mode shall be instigated when the charger is not utilized and a charger status variable is assigned to read the utilization of charger. This variable is then utilized in the mode selection algorithm to select an appropriate mode.

In the first switching state, the switching device S_3 is turned *on* and the inductor current equation is given by

$$\frac{di_{L1}}{dt} = \frac{V_i}{L_1} \tag{5}$$

The storage battery gets charged during the second switching state when the switching device S_3 is turned *off* and the battery charging port switch S_2 is turned *on*. The inductor current discharging slope is given by

$$\frac{di_{L1}}{dt} = \frac{V_i - V_b}{L_1} \tag{6}$$

In these three operating modes, the active switching devices and ports involved are listed in **Table 1** based on which the control variable for closed loop control may be chosen.

2.3 Mode selection

The availability of the multiple modes opens up the possibility of optimal usage of the sources and storage units with minimum cost and user preference by choosing an optimal mode. The primary challenge in a multiport charger is to choose a source depending on the different generation and power supply capability at any given time. Such a source selection algorithm has to measure all the relevant parameters for the source and storage units and decide a specific source based on the available measured data. The source selection should also account for the usage history, energy cost, time of charging, and user preference. A Stateflow-based algorithm is designed to choose an appropriate mode at any given time.

There are three possible modes as listed in **Table 1**. At any time instant, a specific mode has to be chosen based on the system parameters like availability of power, time of charging, and user preference. A flowchart for mode selection and transition is presented in **Figure 4**.

2.3.1 Stateflow-based source selection

Stateflow® is a toolbox available with MATLAB/Simulink that enables one to design state transition tables/diagrams or flowcharts graphically [21]. The mode selection algorithm is developed using the Stateflow tool, and it integrated with the charger modeled with Simulink. Each mode is modeled as a state and the developed diagram is shown in **Figure 5**. The default mode is set to Mode 1 where the solar power is used to charge the vehicular battery. The reference variable for mode selection is the energy demand of the battery ($Ener_d$). At any given instant, the energy demand is compared with the power generated from the source and storage units available in the charger, based on which the mode decision is taken. The third mode, which is idle energy capture mode used to store the generated solar power in the charger battery, needs a separate variable to know whether the charger is engaged with a vehicle or not (*Charger_status*).

Mode	Source port	Load port	Active switching devices	Duty cycle
Mode 1: solar charging mode	V_i	V_O	S_3	D_3
Mode 2: storage charging mode	V_b	V_O	S_1, S_3	D_1, D_3
Mode 3: surplus storage mode	V_i	V_b	S_2, S_3	D_2, D_3 (complementary)

Table 1.
Active switching devices in each mode.

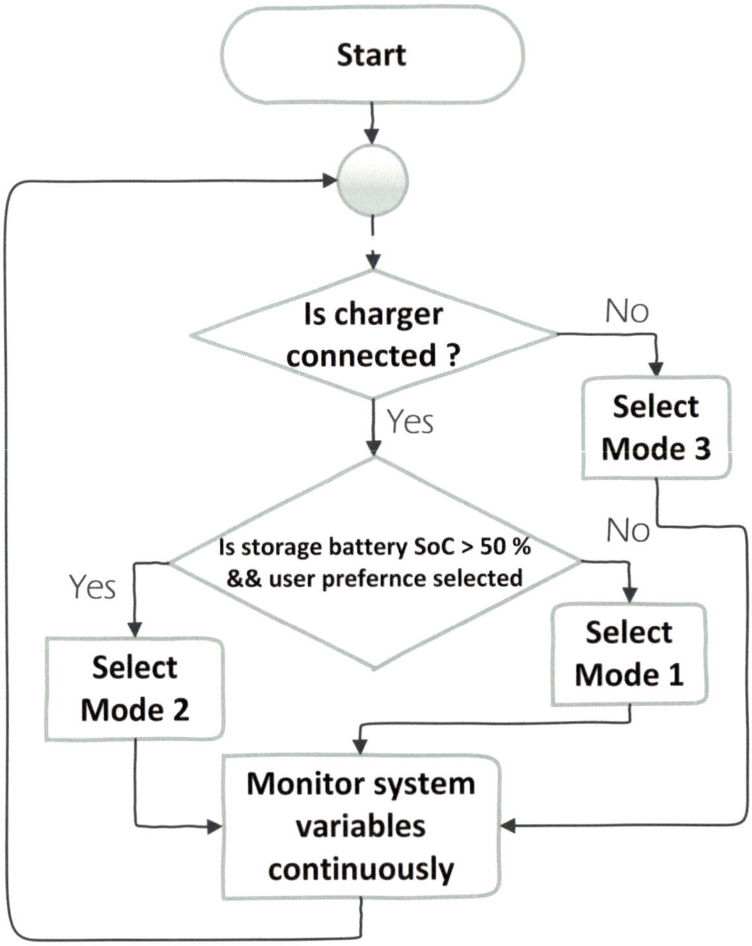

Figure 4.
Flowchart for mode transition.

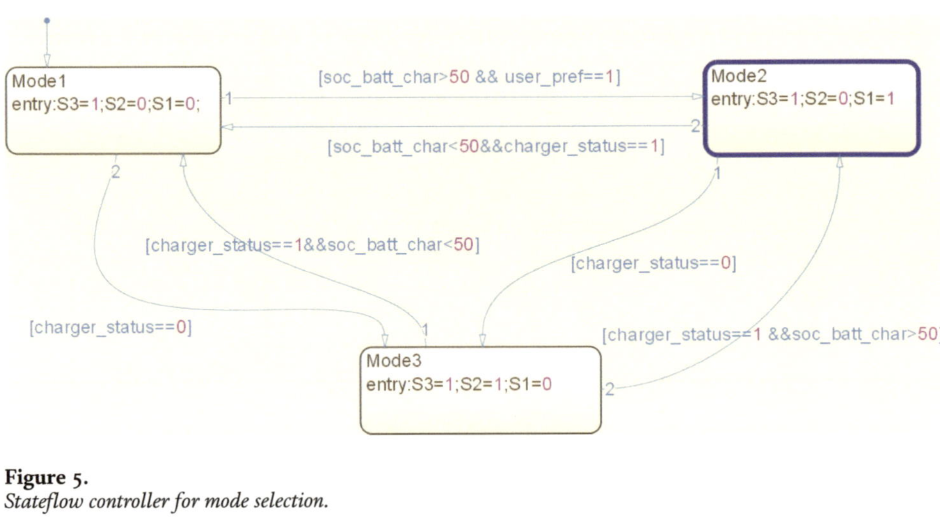

Figure 5.
Stateflow controller for mode selection.

The state of charge (SoC) of the charger battery (*SoC_batt_char*) is actively monitored, and it has to be maintained within the safety limits to ensure the safety of the charger battery. While charging, the SOC of the battery shall not cross the maximum ($SoC\_batt\_char_{max}$), and the minimum limit ($SoC\_batt\_char_{min}$) is monitored while the battery is discharged. A similar condition shall be available for the vehicular battery (*SoC_batt_vehicle*) which would be monitored by the onboard battery management system (BMS).

$$SoC\_batt\_char_{min} < SoC\_batt\_char < SoC\_batt\_char_{max} \tag{7}$$

Considering a case in which both the solar-generated power (P_{solar}) and the available battery power (P_{batt}) are capable of supplying the vehicular battery demand individually, the cost factor (*cost*) is accounted. The charger battery shall have a degradation cost associated with it, which lead to the selection of *Mode* 1 as it is the low-cost supplier at the given time instant. On the other hand, if the user prefers to top up the vehicular battery soon for a ride, the user preference (*user_pref*) is considered, and the source capable of quenching the battery demand is selected.

In case there is no preference set up for the charger, it shall automatically choose the feasible mode based on the measured system variables. While the state chart is running, the user can visually see the active state and the measured data variables that lead to the activation of the current state, as shown in **Figure 5**. The state chart outputs are just Boolean variables and there have to be integrated with the closed loop controller which is shown in **Figure 6**.

2.4 Closed loop controller

In the previous section, the source selection algorithm is covered in detail. Once the source selection is made, the subsequent task is to obtain the regulated power from the available sources. A source selected to charge a vehicular battery unit has to supply regulated power to the batteries irrespective of the time which demands a robust closed loop control algorithm. A predictive control algorithm proposed for a multiport converter to regulate inductor current [22] is considered in this work. The inductor current waveform for the converter in Mode 1 (solar charging mode) shown in **Figure 7** follows a periodic pattern throughout the operation of the converter. The magnitude of the inductor current in the upcoming switching cycle can be precisely predicted with the measured values and duty cycle from the current switching cycle.

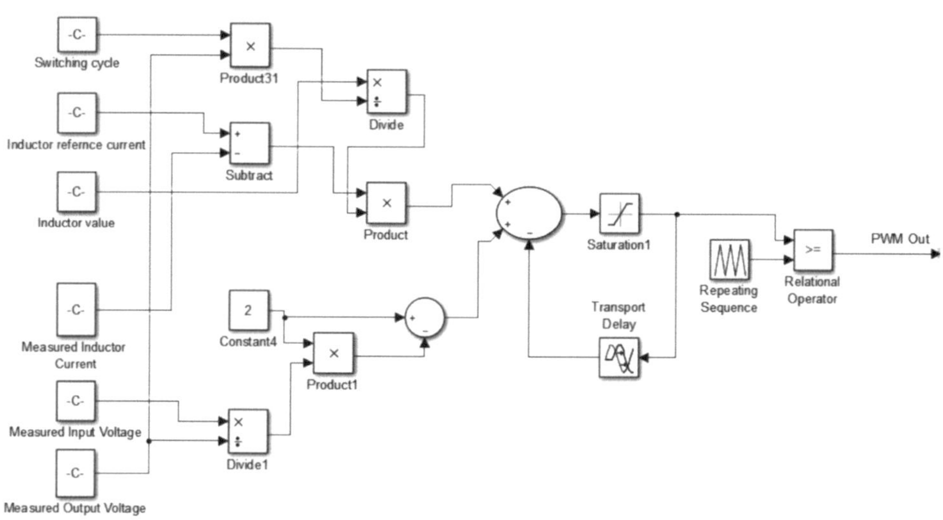

Figure 6.
Structure of predictive current controller.

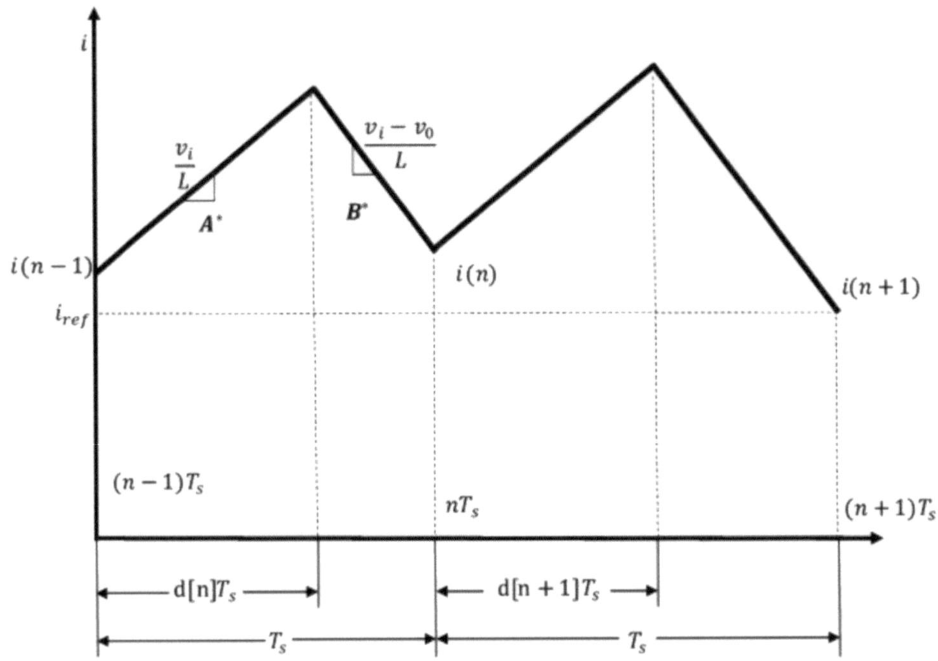

Figure 7.
Inductor current waveform.

Conversely, by adjusting the duty cycle corresponding to the instantaneous value, the inductor current can be made to reach a prefixed value. The method of predicting the duty cycle in the upcoming switching period based on the measurements in the present switching cycle is implemented in predictive controller. The inductor current reaches a prefixed magnitude (i_{ref}) as shown in **Figure** 7 as the measured variable at previous switching instant is utilized for duty cycle prediction.

The converter considered in this work has three different operating modes, and each mode has two switching states. In each of these switching states, the inductor current charge and discharge slopes are derived in the previous section. For Mode 1, the inductor charge and discharge slopes for the nth switching state are shown in **Figure** 7. By knowing the past measured inductor current $i(n-1)$, charge and the discharge slope, the inductor current $i(n)$ can be predicted using Eq. 8 given the duty cycle $d[n]$ is known:

$$i(n) = i(n-1) + \frac{V_i \times d_3[n] \times T_s}{L_1} + \frac{(V_i - V_O) \times d_3'[n] \times T_s}{L_1} \tag{8}$$

The duty cycle summation for a switching cycle considering the on and off period should be unity ($d_3[n] + d_3'[n] = 1$). The prime objective of the controller is to make the inductor current to reach the target current objective i_{ref} within the minimum possible switching states and maintain the same irrespective of the change in the load or source variations. Now, considering the current measured variable $i(n)$, the current in the next switching time instant $i(n+1)$ can be accurately predicted:

$$\begin{aligned} i(n+1) = i(n-1) &+ \frac{V_i \times d_3[n] \times T_s}{L_1} + \frac{(V_i - V_O) \times d_3'[n] \times T_s}{L_1} \\ &+ \frac{V_i \times d_3[n+1] \times T_s}{L_1} + \frac{(V_i - V_O) \times d_3'[n+1] \times T_s}{L_1} \end{aligned} \tag{9}$$

The above equation is rearranged as

$$i(n+1) = i(n-1) + 2\frac{V_i \times T_s}{L_1} - \frac{V_0 \times T_s\left(d_3'[n] + d_3'[n+1]\right)}{L_1} \tag{10}$$

The duty cycle $d_3[n+1]$ is the variable to be computed, and the above equation is rearranged as

$$d_3[n+1] = 2 - d_3[n] + \frac{L_1}{V_0 \times T_s}\{i(n+1) - i(n-1)\} - 2\frac{V_i}{V_0} \tag{11}$$

Conversely, if the current at the next timing instant $i(n+1)$ is set to the reference variable i_{ref}, then the corresponding duty cycle to obtain the i_{ref} can be accurately predicted as $d[n+1]$:

$$d_3[n+1] = 2 - d_3[n] + \frac{L_1}{V_0 \times T_s}\{i_{ref} - i(n-1)\} - 2\frac{V_i}{V_0} \tag{12}$$

The above equation is the control law for predicting the duty cycle in Mode 1. The control law depends on the measurement of the system variables (V_i,V_0). The control law for the other two modes can be obtained by modifying the measured variables relevant to the corresponding mode. The inductor current and the past duty cycle are measured from the corresponding active switch corresponding to each mode listed in **Table 1**. A generic structure for the duty cycle prediction obtained from the control law in Eq. 12 is built in MATLAB/Simulink as shown in **Figure 6**. The mode selection controller decides the mode, and then the predictive controller decides the corresponding duty cycle for the corresponding active switch and thereby the output of the converter is regulated.

3. System modeling and simulation results

3.1 System modeling

3.1.1 Modeling the mode selection controller and charger

The charger topology is modeled using MATLAB/Simulink. The solar PV model from the renewable library is used to model the charger solar generating station. To perform an extensive simulation, the solar irradiance and temperature data is fed into the solar PV model, and the simulation is performed. The charger battery is modeled as a Ni-MH battery, and the vehicle battery is modeled as a Li-ion battery.

3.1.2 Modeling the predictive current controller

The predictive current controller structure is presented in **Figure 6**. The controller designed is a generic one and depends on the measurement of input and output parameters to predict the duty cycle. For instance, in *Mode* 1, the solar PV is available at the input port and the vehicular battery at the output port. Subsequently, in Mode 3, the solar PV is still at the input port and the charger battery is at the output port. The predictive controller depends on the input and output voltage measurements of the converter to predict the duty cycle, and the proper measurements are routed to the corresponding measured variable ports as and when the modes are changed. A look-up table is built from open-loop simulations to identify the inductor current reference under different operating conditions. The predictive current controller is also dependent on the inductor value, and it is assumed that the inductor value remains constant throughout the operation of the converter.

3.1.3 Integrating the predictive current controller with the Stateflow controller

It can be identified from **Figure 8(b)** that the output of the mode selection controller are only boolean variables that help actuate a switching device based on the input and status variables. On the other hand, the predictive current controller generates pulse width modulation (PWM) signals suitable for switching the converter at an appropriate duty cycle to achieve the control target. The Stateflow controller and the predictive current controller are integrated as shown in **Figure 8(b)**. The Boolean mode selection Stateflow output is AND gated with the PWM output from the predictive current controller output. As a result of this, the PWM output from the closed loop controller is directed to the appropriate switching device based on the mode chosen by the Stateflow controller. The Simulink model of the proposed charger topology is shown in **Figure 8(a)**. This model shows the solar PV, charger, and vehicular battery. The gate terminal of switching devices in the charger topology is connected to the controller through "goto" blocks in Simulink.

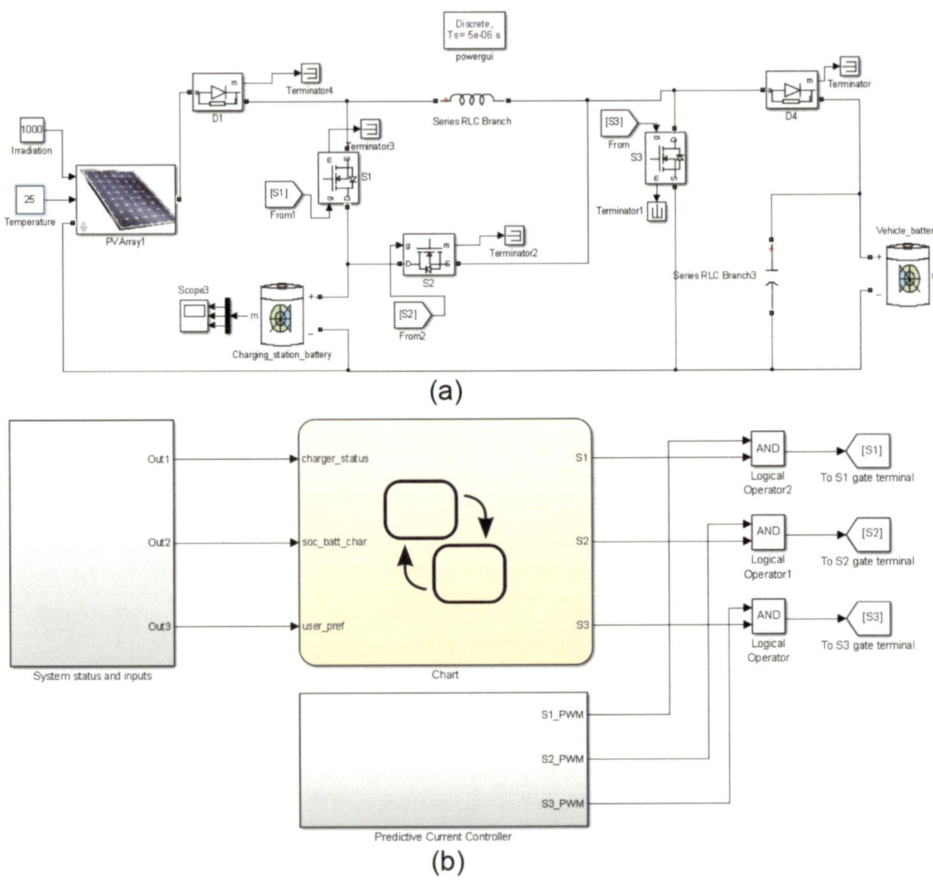

Figure 8.
Integrating the controller with charger. (a) MATLAB/Simulink model of the proposed charger topolog. (b) Mode selection Stateflow controller with predictive current controller.

3.2 MATLAB simulation

The complete charger system thus modeled using MATLAB/Simulink is simulated in parts initially to assess the feasibility of the individual building blocks. The specification of the charger used in simulation is listed in **Table 2**. The input parameters for the Stateflow controllers are modified at first to test the different operating conditions and the corresponding mode election is verified. The different conditions for mode transition are listed in **Table 3** whose variables are defined in Section 2.3.1. For the change in the status variables, manual switches and sliders are used at this stage which is replaced with the actual user input and status variables from the modeled system at the later stage while performing system level simulation.

Similarly, the predictive current controller is tested by applying a load step variation at the load port. The emulated load step variation using the resistance of different values shall make the load current to change, but the predictive current controller modifies the duty cycle to make the inductor current constant irrespective of the load conditions. Once these individual simulations are done, the model has to be modified to run the system level simulation.

3.2.1 Real-time simulation using OPAL-RT

A simulation model represents the physical behavior of a system through the operation or use of another. With the advent of mathematical modeling and digital tools, digital simulation has become prevalent. In any discrete-time simulation, a set of equations are solved at every time step which may be fixed or variable. Specifically, in a fixed-step simulation that is running on a generic purpose computer, the fixed time step taken to compute the system of equations may be longer or shorter than the actual time step.

On the other hand, the real-time simulation required must perform similar computation with the duration similar to the physical world [23]. For simulating nonlinear systems like power electronic systems, the change in the actual time step

Primary source	**Solar PV (2 kWp)**
Storage	Battery (20 kWh)
Vehicular battery	Li-manganese (16 kWh)
Maximum charging current	15 A
Type	Level 1 DC charger [19]

Table 2.
Simulation parameters.

S. no	**Mode 1**	**Mode 2**	**Mode 3**
Mode 1	—	[soc_batt_char>50 && user_pref==1]	[charger_status==0]
Mode 2	[soc_batt_char <50&&charger_status==1]	—	[charger_status==0]
Mode 3	[charger_status==1&&soc_batt_char <50]	[charger_status==1 &&soc_batt_char>50]	—

Table 3.
Mode transition conditions.

may lead to erroneous or inaccurate results. Hence, performing a real-time simulation with small time steps help the simulation to represent the physical behavior of nonlinear systems accurately. Among the different real-time simulators, OPAL-RT is an FPGA platform that supports real-time simulation and is fully integrated with MATLAB/Simulink.

3.2.2 Preparing MATLAB model for real-time simulation

The OPAL-RT real-time simulation requires that the whole MATLAB/Simulink model be made into two subsystems, namely, master and console. It should be noted that the "powergui" block should be placed on the top model and not inside any subsystems. Initially, the model is prepared for RT-LAB simulation which is later used for real-time simulation. The steps for conversion are shown in **Figure 9 (a)**. As specified in the previous section, the simulation should be run using a discrete fixed-step time solver. The time step should be carefully chosen and the other simulation parameters related to time should be an integral multiple of the time step. Additionally, in the MATLAB settings under the "Model configuration parameter," the block reduction settings should be turned off. The optimization and signal reuse should be turned off. Once the model is prepared, a free run should be done with the simulation time set to infinity (inf).

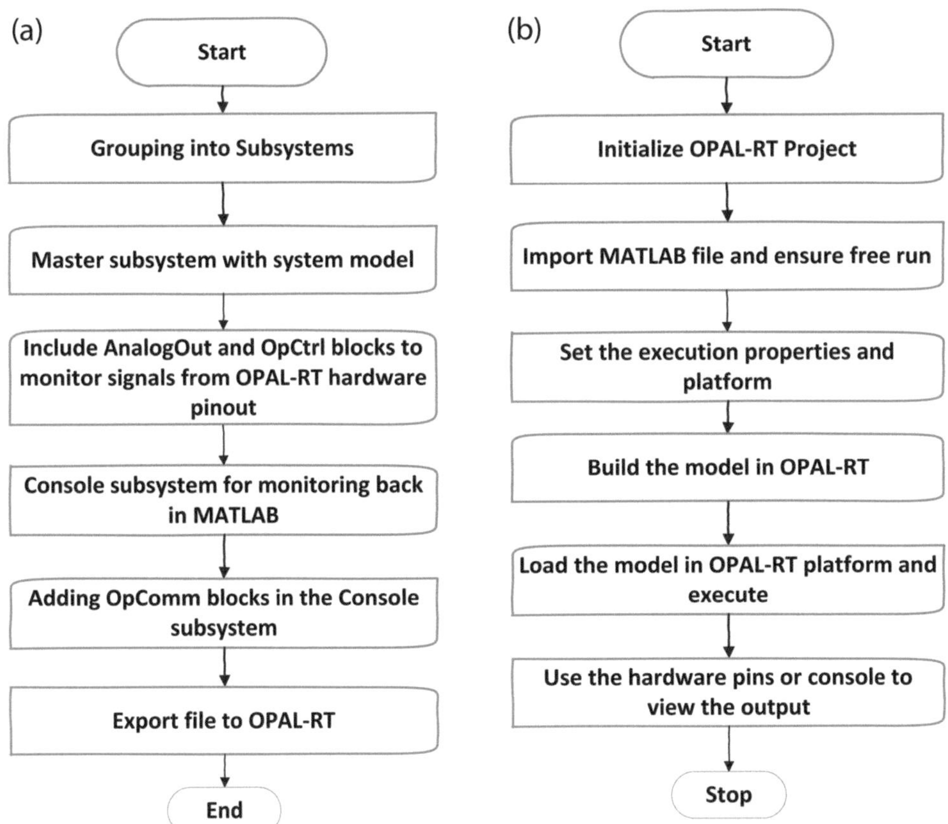

Figure 9.
Steps to perform real-time simulation. (a) Steps for preparing MATLAB model for RT simulation. (b) Steps to run real-time simulation using OPAL-RT.

3.2.3 Master subsystem

The master subsystem shall be named with a prefix "SM_" which is an identifier for the OPAL-RT to identify what has to be taken to OPAL-RT simulator. Apart from the MATLAB built model, the "OpCtrl" block has to be included in the model. This enables the pin configuration of the OPAL-RT to be flashed into the simulator, and the configuration file should be placed in the project folder. For every signal whose monitoring is required from OPAL-RT platform, an "AnalogOut" block is added. The analog outputs of the OP4500 can produce only up to 5 V; hence all signals should be accompanied along with a suitable multiplier to scale up/down the signal.

3.2.4 Console subsystem

The console subsystem is intended for acquiring and monitoring the signals from the OPAL-RT platform back to MATLAB and to view the signals in MATLAB scope. Each signal entering into this subsystem shall be prefixed with an "OpComm" block which shall help in matching the fetching rates of the MATLAB computer along with the OPAL-RT platform. The final prepared model is shown in **Figure 10**.

3.3 Simulation results

3.3.1 Simulating in OPAL-RT platform

The MATLAB model is now ready to be simulated with the OPAL-RT platform. The steps for performing the real-time simulation on OPAL-RT platform are shown in **Figure 9(b)**. The developed model is imported in the OPAL-RT tool, and the model is built which is then loaded into the OPAL-RT platform for real-time simulation. The proposed charger topology has to be tested for regulated output and mode selection. The load step variation is induced on the load port, and the

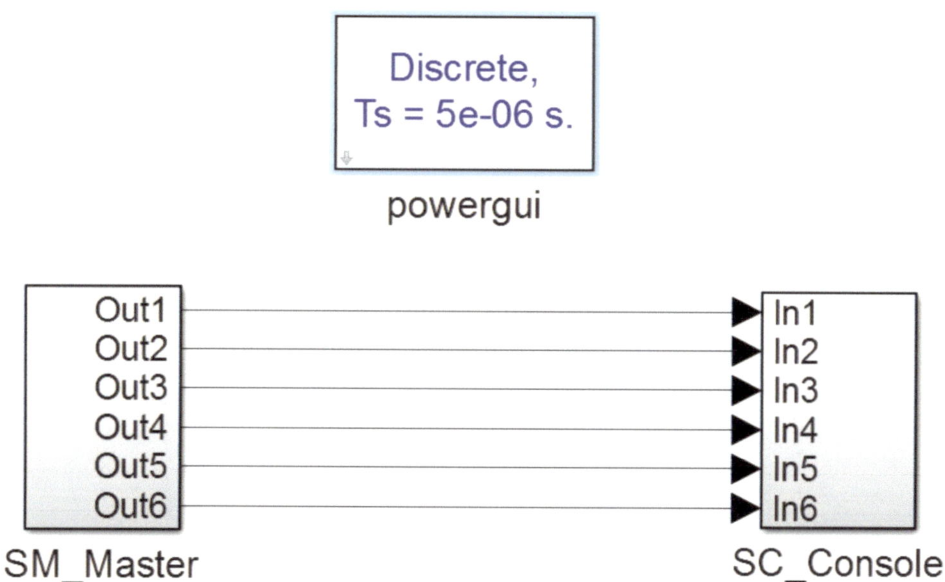

Figure 10.
MATLAB/Simulink for real-time simulation.

variations the inductor current and the output voltage are presented in **Figure 11**. It can be observed that the predictive current controller regulates the inductor current within the prefixed limits.

The charger power profile is shown in **Figure 12** which gives the 24-hour charger power output. A typical household with an overnight slow charging is considered, and the EV is not available for charging from 09.00 to 19.00 hours. The charger power output is capped at 1 kW, and depending on the power available from the sources and the load requirement, different modes are instantiated.

3.3.2 Viewing results from MATLAB console

Once the OPAL-RT model is executed, a system generated MATLAB console opens, which is used to monitor and log the signals running from the OPAL-RT platform. The effectiveness of the predictive current controller is observed by introducing a load step variation. The current controller tightly regulates the current irrespective of the load step at 0.025 s, and the results are shown in **Figure 11**.

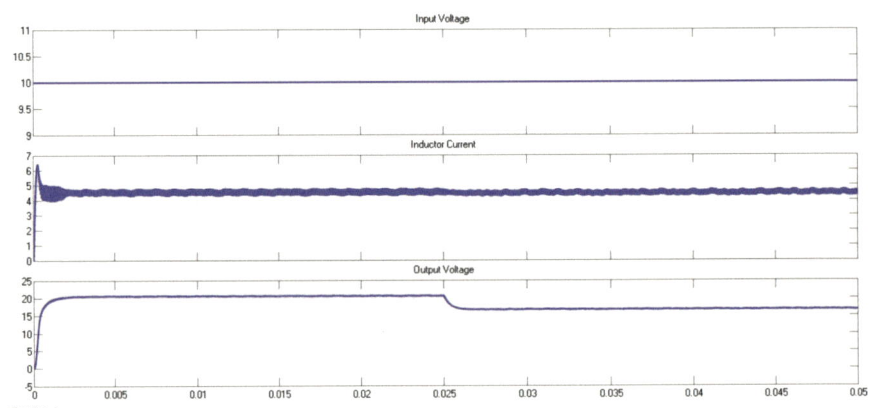

Figure 11.
Simulation results of load step variation observed on console.

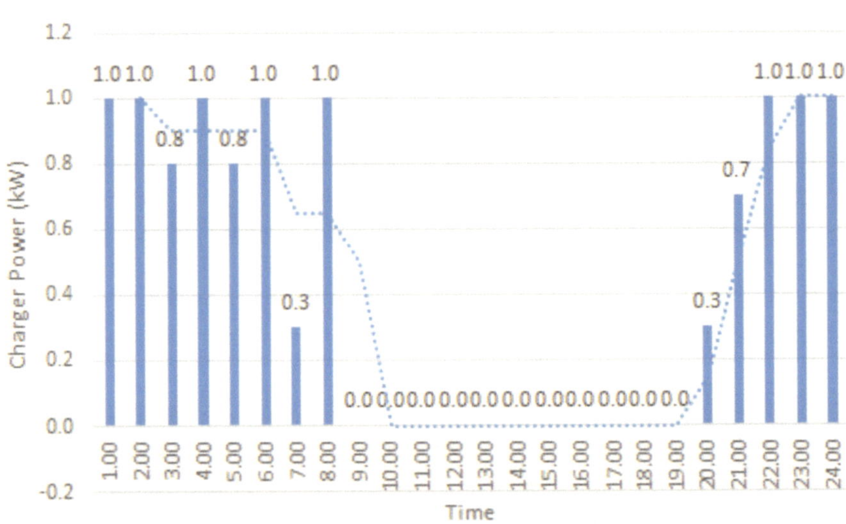

Figure 12.
Charger power profile [24].

3.3.3 Viewing results from hardware pinout

In addition to the MATLAB console, the model output can be directly read from the hardware pins available in OPAL-RT platform. This work was performed on an OP4500 platform that supports up to 96 input/output channels. The MATLAB running computer is interfaced with the OP4500 using a 5-Giga bit optical fiber cable, and the real-time simulation is performed on a XILINX Kintex 7 FPGA present inside the OP4500. A typical steady-state condition corresponding to *Mode* 2 is captured in **Figure 13** which shows the PWM outputs of the two switching devices in the charger. Subsequently, the Stateflow controller is tested by changing the *charger_status* variable. The current mode is *Mode* 2 as highlighted in

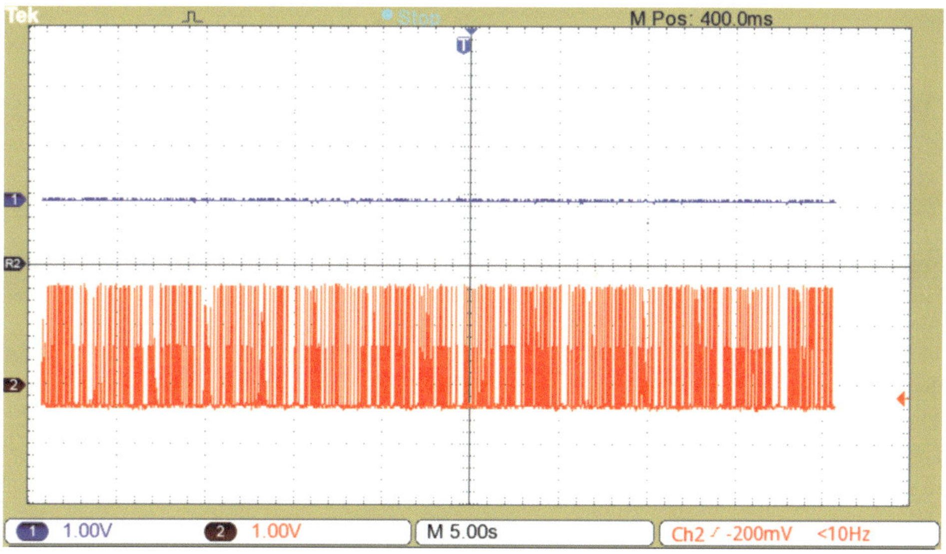

Figure 13.
Steady-state results of the Stateflow controller [VGS1 on CH1 and VGS2 on CH2].

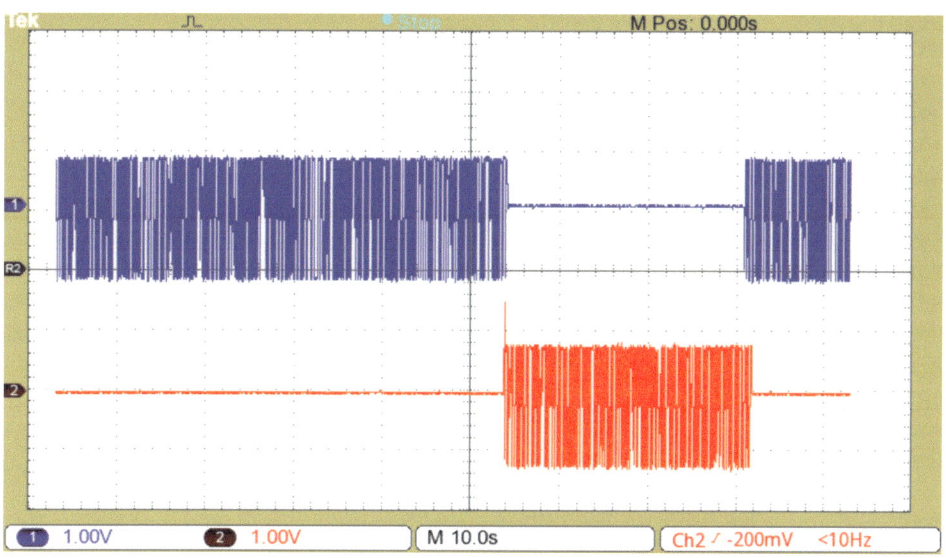

Figure 14.
Gate pulses during mode transition [VGS1 on CH1 and VGS2 on CH2].

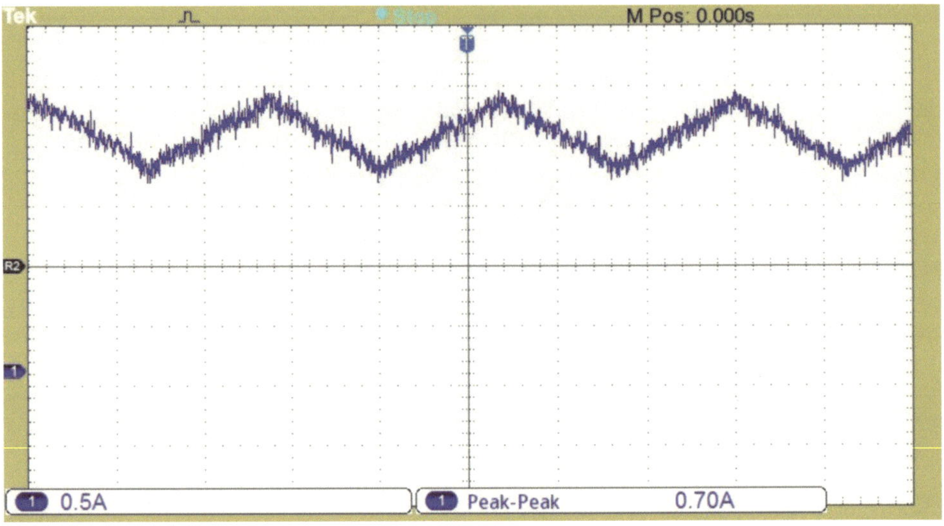

Figure 15.
Inductor current during normal operating conditions.

Figure 5 and the change in the status of this variable shall change the active mode to *Mode* 3 based on the conditions listed in **Table 3**. The mode transition results in the change in the switching pulse signals as shown in **Figure 14**.

The Stateflow controller is an efficient method to monitor the state variables and the system parameters which could be used to choose an appropriate mode based upon the operating conditions. The inductor current waveform during the normal operating conditions follows a typical charge and discharge pattern as shown in **Figure 15**.

4. Conclusion

This work present the real-time simulation of a three-port electric vehicle charger with solar PV assisted battery as energy sources. With three different operating modes, the prime objective of the energy management controller is to choose an appropriate mode based on the system parameters. A Stateflow-based mode selection controller is adopted, and its development is briefed in this chapter. Once the mode selection is done, the secondary function of the energy management controller is to regulate the system variable and obtain the control objective irrespective of the load variations. A predictive current controller is introduced and the control law for duty cycle prediction is derived. The two individual controllers for mode selection and output regulation are integrated to form the energy management algorithm for the three-port EV charger. A MATLAB/Simulink model is developed to simulate the proposed charger topology and simulated. Finally, the developed charger model is loaded into the OPAL-RT platform, and real-time simulation is performed. The methodology proposed in this chapter shall be extended to any charger topology, and the steps explained in this chapter may be useful in performing real-time simulation of nonlinear systems. The major limitation of this charger topology is lack of bidirectional power flow. By changing the topology suitable for vehicle-to-grid, the battery power from the vehicle could be used for grid load demand reduction and peak shaving in load profiles. The future work in this domain shall integrate the charting infrastructure with the information communication technology for efficient power delivery and monitoring.

Acknowledgements

The authors wish to acknowledge the support provided by Electric Vehicle Engineering and Robotics (EVER Labs) Laboratory at SASTRA Deemed University, Thanjavur, India, to utilize the OPAL-RT platform for performing experiments related to this work. The support provided by Mr. Venkatavasan is instrumental in running the experiments on OPAL-RT platform. The authors would like to express their heartfelt gratitude for the sponsorship provided by IntechOpen and Knowledge Unlatched, which covered the open-access publication fee for this chapter.

Nomenclature

i_{L1}	inductor current
V_i	input voltage
L_1	inductor
V_0	output voltage
V_b	battery voltage
S_1, S_2, S_3	switching devices 1, 2, and 3
D_1, D_2, D_3	duty cycle of the switching devices 1, 2, and 3
SoC_batt_char	state of charge (SoC) of the charger battery
$SoC\_batt\_char_{max}$	maximum state of charge (SoC) of the charger battery
SoC_batt_vehicle	state of charge (SoC) vehicular battery
user_pref	user preference for mode selection
i_{ref}	reference inductor current
$i(n-1)$	inductor current during the previous switching cycle
$i(n+1)$	inductor current during the next switching cycle
T_s	switching time
$d[n]$	duty cycle during the present switching cycle

Author details

Santhosh Thuttampatty Krishnamoorthy[1*], Suthanthira Vanitha Narayanan[2] and Ramkumar Kannan[1]

1 Centre for Energy Storage and Conversion, School of Electrical and Electronics Engineering, Shanmuga Arts Science Technology and Research Academy, Thanjavur, India

2 Department of Electrical and Electronics Engineering, Muthayammal Engineering College, Tamil Nadu, India

*Address all correspondence to: santhosh@eee.sastra.edu

References

[1] Global EV Outlook 2018. 2018. DOI: 10.1787/9789264302365-en

[2] Zehner O. Unclean at any speed. IEEE Spectrum. 2013;**50**:40-45. DOI: 10.1109/MSPEC.2013.6545121

[3] Voelcker J. Electric vehicles need more study, less emotion. IEEE Spectrum. 2013;**50**:8. DOI: 10.1109/MSPEC.2013.6565543

[4] Juyal S, Sanjeevi H, Saxena A, Sharma S, Singh A, Chander S, et al. Zero Emission Vehicles (ZEVs): Towards a Policy Framework. 2018

[5] Society of Indian Automobile Manufacturers. White Paper on Electric Vehicles Adopting Pure Electric Vehicles: Key Policy Enablers; 2017. p. 32

[6] Chakraborty S, Vu H-N, Hasan MM, Tran D-D, Baghdadi, El M, Hegazy O. DC-DC converter topologies for electric vehicles, plug-in hybrid electric vehicles and fast charging stations: State of the art and future trends. Energies. 2019;**12**: 1569. DOI: 10.3390/en12081569

[7] Pillai RK, Suri R, Kundu S, Singh H, Sarkar Roy S, Dhuri S. Electric vehicle charging stations business models for India - ISGF Report. 2018

[8] Gallardo-Lozano J, Milanés-Montero MI, Guerrero-Martínez MA, Romero-Cadaval E. Electric vehicle battery charger for smart grids. Electric Power Systems Research. 2012;**90**:18-29. DOI: 10.1016/j.epsr.2012.03.015.

[9] Khaligh A, DAntonio M. Global trends in high-power on-board chargers for electric vehicles. IEEE Transactions on Vehicular Technology. 2019;**9545**:1. DOI: 10.1109/tvt.2019.2897050

[10] Ahmad A, Alam MS, Chabaan R. A comprehensive review of wireless charging technologies for electric vehicles. IEEE Transactions on Transportation Electrification. 2017;**4**: 38-63. DOI: 10.1109/TTE.2017.2771619

[11] Liu Z, Song Z. Robust planning of dynamic wireless charging infrastructure for battery electric buses. Transportation Research Part C: Emerging Technologies. 2017;**83**:77-103. DOI: 10.1016/j.trc.2017.07.013

[12] Sujitha N, Krithiga S. RES based EV battery charging system: A review. Renewable and Sustainable Energy Reviews. 2016;**75**:978-988. DOI: 10.1016/j.rser.2016.11.078

[13] Khan SA, Islam MR, Guo Y, Zhu J. A new isolated multi-port converter with multi-directional power flow capabilities for smart electric vehicle charging stations. IEEE Transactions on Applied Superconductivity. 2019;**29**:1-4. DOI: 10.1109/TASC.2019.2895526

[14] Sedaghati R, Shakarami MR. A novel control strategy and power management of hybrid PV/FC/SC/ battery renewable power system-based grid-connected microgrid. Sustainable Cities and Society. 2019;**44**:830-843. DOI: 10.1016/j.scs.2018.11.014

[15] Alam MN, Chakrabarti S, Ghosh A. Networked microgrids: State-of-the-art and future prospectives. IEEE Transactions on Industrial Informatics. 2018;**15**:1238-1250. DOI: 10.1109/ TII.2018.2881540

[16] Liu Y, Li Y, Liang H, He J, Cui H. Energy routing control strategy for integrated microgrids including photovoltaic, battery-energy storage and electric vehicles. Energies. 2019;**12**: 302. DOI: 10.3390/en12020302

[17] Petreus D, Etz R, Patarau T, Cirstea M. An islanded microgrid energy management controller validated by

using hardware-in-the-loop emulators. International Journal of Electrical Power & Energy Systems. 2019;**106**:346-357. DOI: 10.1016/j.ijepes.2018.10.020

[18] Santhosh TK, Natarajan K, Govindaraju C. Synthesis and implementation of a multi-port DC/DC converter for hybrid electric vehicles. Journal of Power Electronics. 2015;**15**: 1178-1189. DOI: 10.6113/JPE.2015.15. 5.1178

[19] Yilmaz M, Krein PT. Review of battery charger topologies, charging power levels, and infrastructure for plug-in electric and hybrid vehicles. IEEE Transactions on Power Electronics. 2013;**28**:2151-2169. DOI: 10.1109/TPEL.2012.2212917

[20] Kanchev H, Lu D, Colas F, Lazarov V, Francois B. Energy management and operational planning of a microgrid with a PV-based active generator for smart grid applications. IEEE Transactions on Industrial Electronics. 2011;**58**: 4583-4592. DOI: 10.1109/TIE.2011. 2119451

[21] MathWorks. Stateflow Documentation. 2017. Available from: https://in.mathworks.com/help/statef low/index.html

[22] T K S, C G. Development of predictive current controller for multi-port DC/DC converter. International Journal of Power Electronics and Drive Systems. 2015;**6**:683. DOI: 10.11591/ ijpeds.v6.i4.pp683-692

[23] Belanger J, Venne P, Paquin J-N. The what, where and why of real-time simulation. Planet Rt. 2010;**1**:25-29

[24] Jian L, Xue H, Xu G, Zhu X, Zhao D, Shao ZY. Regulated charging of plug-in hybrid electric vehicles for minimizing load variance in household smart microgrid. IEEE Transactions on Industrial Electronics. 2013;**60**:3218-3226. DOI: 10.1109/TIE.2012.2198037

Chapter 7

Modelling and Control of Narrow Tilting Vehicle for Future Transportation System

Yaxing Ren

Abstract

The increasing number of cars leads traffic congestion and parking problems in urban area. Small electric four-wheeled narrow tilting vehicles (NTV) have the potential to become the next generation of city cars. However, due to its narrow width, the NTV has to lean into corners like two-wheeled vehicles during a turn. It is a challenge to maintain its roll stability to protect it from falling down. This chapter aims to describe the development of NTV and drive assistance technologies in helping to improve the stability of an NTV in turning. The modelling of an NTV considers the dynamics of the tyres and power train of the vehicle. A nonlinear tilting controller for the direct tilting control mechanism is designed to reduce the nonlinear behaviour of an NTV operating at different vehicle velocities. In addition, two torque vectoring based torque controllers are designed to reduce the counter-steering process and improve the stability of the NTV when it turns into a corner. The results indicate that the designed controllers have the ability to reduce the yaw rate tracking error and maximum roll rate. Then riders can drive an NTV easily with the drive assistance system.

Keywords: modelling, stability control, nonlinear control, torque vectoring, drive assistance system, narrow tilting vehicle

1. Background

Vehicle is one of the most widely used transportation in people's daily life. Due to the greenhouse gas emission problem of traditional cars, the development of electric vehicles received great attention in recent years. However, in urban area, the increasing number of cars causes the traffic congestion problems and limit parking places. Because of these issues, small narrow commuter vehicles are expected to become a new generation of city cars [1, 2]. In the UK, some researches have been done in the past years, such as the CLEVER Project [3–5] and RESOLVE Project [6, 7], which developed two prototype vehicles as shown in **Figure 1** [8]. This kind of vehicles are also called narrow tilting vehicles (NTVs) have just half the width of a conventional car like a motorcycle but have four wheels like a car. This makes an NTV a convergence of a car and a motorcycle that makes it integrate the features and advantages of them.

The conventional four-wheel vehicles is wide that have enough roll stiffness to balance the roll stability by its own suspension structure. But the NTV has no such roll stiffness and have to lean into corners during turning [3, 9], as shown in

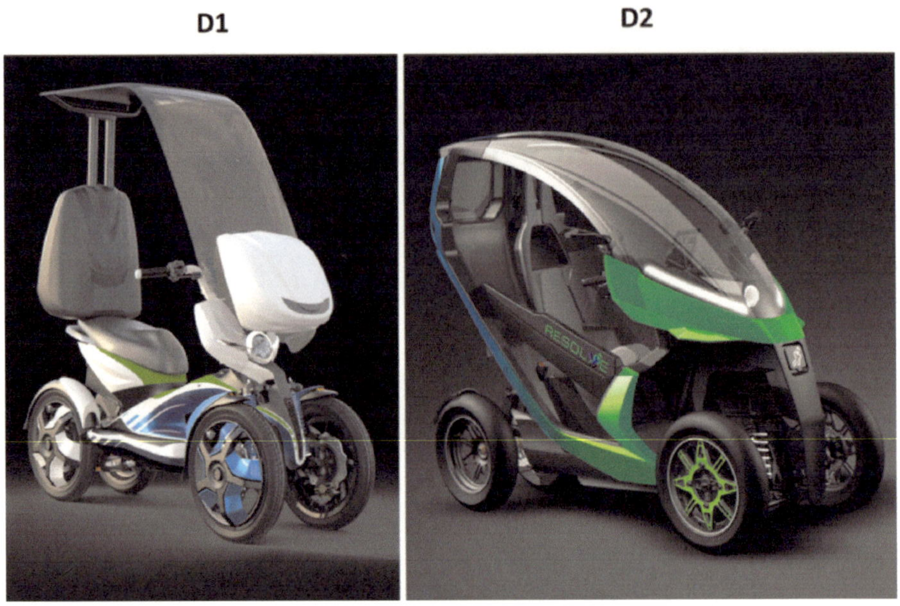

Figure 1.
Two demonstrators of narrow tilting vehicle developed in the RESOLVE Project [8].

Figure 2.
The diagram of narrow tilting vehicle in RESOLVE Project developed by AIT [6].

Figure 2. This is a challenge to the roll stability of NTV and need to be maintained by experienced riders. The purpose of the vehicle design is not only driven by experienced riders but also new riders. Thus, the autonomous drive assistance system is required to improve its roll stability [10–12].

In riding a motorcycle, the rider can lean the motorcycle into a corner by shifting his own weight. But this is not available in riding an NTV as the weight of a human body is much lighter than that of the vehicle. Thus, in riding an NTV, the rider has to act on the throttle with counter-steering process to balance the vehicle in a turn [3, 9]:

- the rider provides a counter-steering on the throttle on an opposite direction;

- the counter-steering provides the opposite lateral force;

- the lateral force rolls the vehicle into the expected side;
- the rider then turns the steering back to the expected direction at an appropriate moment; and
- the vehicle stops rolling down and yaws to the expected route.

It shows that the riders of NTVs have to be very experienced in balancing the vehicle and following the path simultaneously. The NTV can be in different tilting states, such as straight on, turning, accelerating turn, oversteer/understeer and highsider, as shown in **Figure 3**. However, the next generation vehicles are expected to be easy-driving to low-experienced riders. A drive assistance system can help new riders in balancing the vehicle and the riders only need to focus on the path in the riding. This leads the development of an autonomous drive assistance system for NTV.

To improve the tilting stability, the common solution is to design the active tilting control via installing additional mechanisms. The two main tilting methods are the steering tilt control (STC) and the direct tilt control (DTC) on different mechanisms [13, 14], as shown in **Figure 4**. The STC directly controls the steering angle of front wheel to autonomously complete the tilting process as an experienced rider for stabilising the vehicles, while the DTC provides additional torque to lean the vehicle to the expected corners.

The STC system is efficient at high speed but performs worse at the standstill or very low speeds. In slippery road conditions, the performance of using STC is even worse [15]. The DTC based mechanism slightly simplifies these control problems with an additional control input from a separate tilt actuator [16]. But the DTC

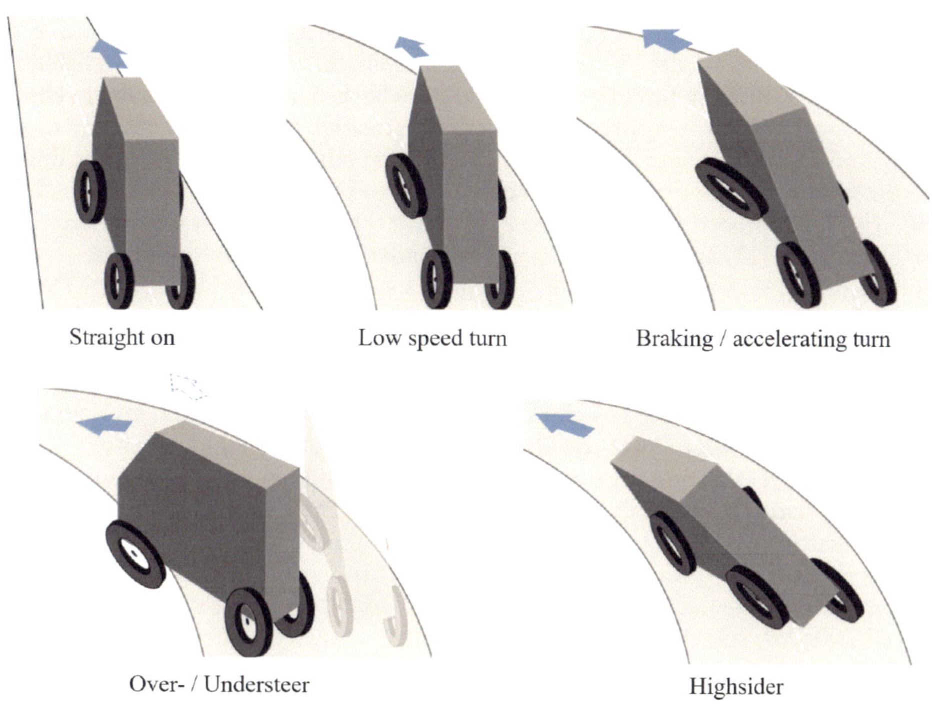

Figure 3.
Tilting states of NTV.

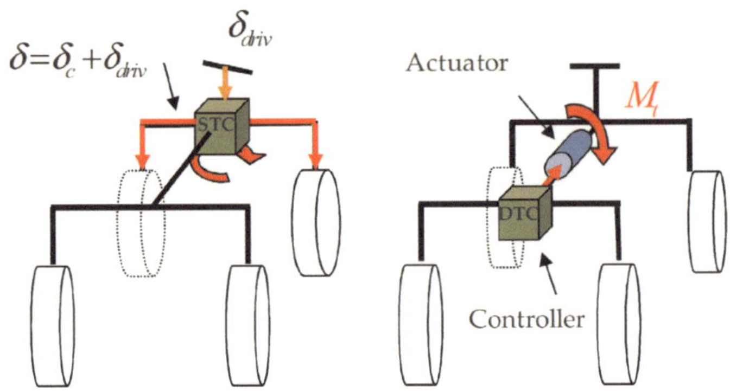

Figure 4.
The STC and DTC tilting mechanisms of NTV [10].

system requires high tilting motion at high vehicle speed and has risk to cause the vehicle oscillations. In addition, the delayed vehicle response speed could reduce the performance of tilting motion. Thus, it requires highly sophisticated loop control algorithms adapting to different loads and driving conditions [16]. The combination of STC and DTC in dual mode switching strategies is available to reduce their drawbacks. But such approaches have obvious discontinuous behaviour during the mode switching [4, 14, 17].

Several studies focused on the control approach design in driving the DTC actuator, including linear SISO control approaches to provide tilt torque from a given combination of vehicle information [2, 18], model-based control methods to decouple the longitudinal and lateral dynamics in vehicle response [5, 19, 20], and nonlinear control solutions to compensate the nonlinear behaviour of NTVs based on the accurate vehicle model [20, 21].

On the other side, the torque vectoring (TV) technology is able to improve the vehicle cornering response and potential to improve the handling performance of a vehicle [22]. The left-right TV technique was proposed in [23] aiming to distribute the driving and braking forces on left and right wheels in a wheel-individual vehicle. The control allocation criteria were verified sensitive to electric motor drive parameters in performance comparison [24]. The maximum vectoring torque limit was determined in [25] and desired traction force and yaw moment were mapped in [26]. The TV approach was optimised to improve the yaw moment distraction performance in [27] and the stability of NTV under expected environmental conditions in [28]; the lateral stability in cornering was enhanced by optimal TV approach to maximise the vehicle velocity in [29]; and minimised the power losses of TV to improve the battery efficiency [30].

In these approaches, the TV method is used as an assistant torque mainly for improving the performance of vehicle yaw turn and enhancing the lateral stability. The yaw moment on a vehicle can also affect the roll stability and it is more sensitive to an NTV. The conventional TV methods and their optimisation may not suitable for both yaw and roll stability enhancement. Thus, the roll stability maintenance of using the TV technology needs to be paid more attention in an NTV.

This chapter first designs a nonlinear tilting controller for DTC-based NTVs without the dependence of an accurate vehicle model to improve the performance of DTC from low speed to high speed. In addition, this chapter develops the TV technology based drive assistance system to maintain the roll dynamics of NTV in cornering. Both approaches are developed to assist the rider in turning an NTV and improve the roll stability of the vehicle. As a result, both the new rider and experienced rider can drive the NTV easily.

2. Mathematical model of four-wheel vehicle dynamics

The basic model of NTV was proposed in 1990s for the two or three wheeled tilting vehicles from the simplified bicycle geometric model [17, 31, 32]. The model considers the vehicle body dynamic only and the wheels are assumed with light weighting and not leaning with the vehicle body. The University of Minnesota proposed the nonlinear NTV model with wheel dynamics considering the distribution force on each wheel [16, 20, 33]. The University of Bath proposed a five DoF nonlinear model of NTV and wheel dynamics [3–5]. On the other hand, due to the contacts between road surface and tyre significantly affects the friction force of wheel dynamics, the tyre longitudinal slip can be combined into the wheel dynamic in the NTV model. This section will discuss the detailed NTV model to describe the dynamics of traction force transferred from tyre to vehicle body and simplified single-track vehicle model for controller design.

2.1 Wheel dynamics

As seen in the wheel dynamic model in **Figure 5**, the wheel speed ω_{ij} describes the power transfer from rear-wheel-drive wheel hub to road. In four-wheeled vehicle, the wheels of front left, front right, rear left, and rear right wheels are represented as $ij \in \{\text{fl}, \text{fr}, \text{rl}, \text{rr}\}$. The traction torque T_{rj} is applied on rear left and rear right wheels and the brake torque $T_{\text{brk},i}$ are applied on the centre of all wheels. The longitudinal force $F_{\text{l},ij}$ is the force to drive the wheels at the contact point between tyre and road surface.

The dynamics of wheel speeds are represented as [12, 34]:

$$\dot{\omega}_{\text{f}j} = \frac{-T_{\text{brk,f}} - R_{\text{f}} F_{\text{l,f}j}}{J_{\text{f}j}} \tag{1}$$

$$\dot{\omega}_{\text{r}j} = \frac{T_{\text{r}j} - T_{\text{brk,r}} - R_{\text{r}} F_{\text{l,r}j}}{J_{\text{r}j}} \tag{2}$$

where J_{ij} is the wheels' inertia around the wheel with the radius R_i.

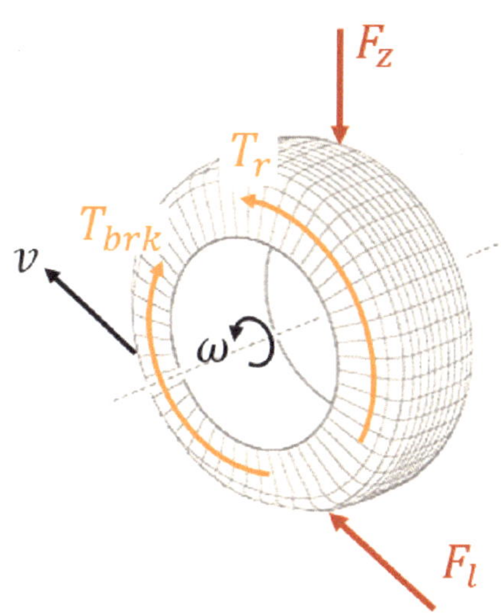

Figure 5.
Wheel dynamic model.

The longitudinal force can be described as a function of friction coefficient μ_{ij} and tyre longitudinal slip $s_{l,ij}$ as

$$F_{l,ij} = F_{z,ij} \cdot \mu_{ij}\left(s_{l,ij}\right) \tag{3}$$

where the tyre longitudinal slip $s_{l,ij}$ can be described based on the vehicle velocity v and vehicle side-slip angle β as:

$$s_{l,ij} = \frac{R_i\omega_{ij} - v\cos\beta}{\max\left(R_i\omega_{ij}, v\cos\beta\right)} \tag{4}$$

which describes the longitudinal slip of both acceleration and braking conditions, as shown in **Figure 6**.

The tyre characteristics of friction coefficient μ_{ij} are modelled by the *magic tyre formula* as [35].

$$\mu_{ij}\left(x_{ij}\right) = D \cdot \sin\left\{C\arctan\left[B(1-E)\cdot x_{ij} + E\arctan\left(B\cdot x_{ij}\right)\right]\right\} \tag{5}$$

where B, C, D and E are the parameters to determine the friction coefficient of tyre, the input x_{ij} can be either longitudinal slip $s_{l,ij}$ to calculate the longitudinal slip friction coefficient or lateral slip angle α_{ij} to calculate the side slip friction coefficient [36]. An example of the longitudinal tyre characteristics of friction coefficient with wheel slip ratio on typical roads of dry, wet, snow and iced conditions, as shown in **Figure 7**.

The vertical load $F_{z,ij}$ of each wheel can be calculated by

$$F_{z,fl} = m\left(\frac{l_r}{l}g - \frac{h}{l}a_x\right)\left(\frac{1}{2} - \frac{h}{b_f}\frac{a_y}{g}\right) \tag{6}$$

$$F_{z,fr} = m\left(\frac{l_r}{l}g - \frac{h}{l}a_x\right)\left(\frac{1}{2} + \frac{h}{b_f}\frac{a_y}{g}\right) \tag{7}$$

$$F_{z,rl} = m\left(\frac{l_f}{l}g + \frac{h}{l}a_x\right)\left(\frac{1}{2} - \frac{h}{b_r}\frac{a_y}{g}\right) \tag{8}$$

$$F_{z,rr} = m\left(\frac{l_f}{l}g + \frac{h}{l}a_x\right)\left(\frac{1}{2} + \frac{h}{b_r}\frac{a_y}{g}\right) \tag{9}$$

where m is the lumped mass of vehicle itself and rider; g is the gravitational constant; l is the distance of wheelbases consisting l_f and l_r, which represent the distance from the centre of gravity (COG) to front axles and rear axles, respectively; h indicates the vehicle height that is measured from the road surface to the COG of the vehicle; b_f and b_r are the track of front and rear axle; a_x and a_y are the acceleration of vehicle in x-axis and y-axis of the vehicle-fixed coordinate system.

The tyre sideslip force is presented by the magic tyre formula in (5) as

$$F_{s,ij} = F_{z,ij} \cdot \left[\mu_{ij}\left(\alpha_{ij}\right) + \lambda_{st}\theta\right] \tag{10}$$

where θ is the lean angle of wheels, which is the same as lean angle of vehicle, and λ_{st} is the camber stiffness coefficient of wheels. The wheel lateral slip angle α_{ij} represents the angle between the longitudinal axis of wheel and its velocity forward direction. The lateral slip angle can be presented as

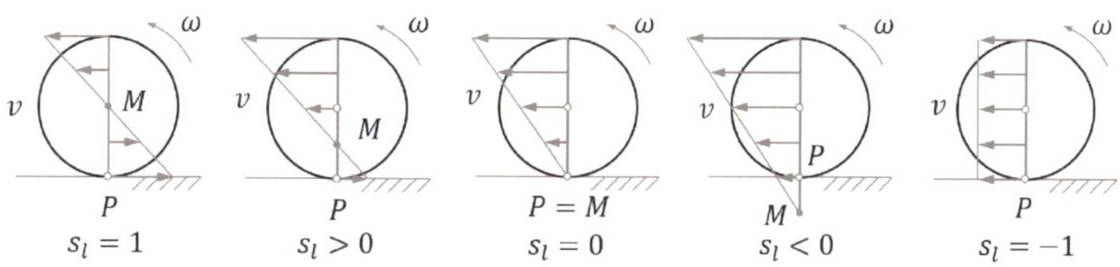

Figure 6.
Wheel slip ratio at different wheel rotating conditions.

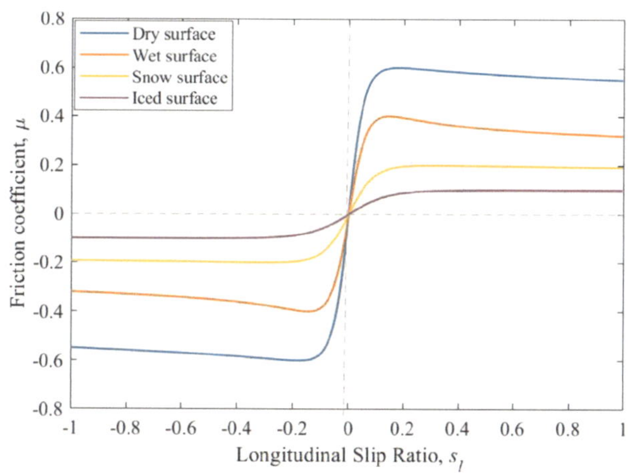

Figure 7.
Longitudinal tyre characteristics of variable friction coefficient with longitudinal slip ratio.

$$\alpha_{\mathrm{f}j} = \delta - \arctan\left(\frac{v\sin\beta + l_{\mathrm{f}}\dot{\varphi}}{v\cos\beta}\right) \tag{11}$$

$$\alpha_{\mathrm{r}j} = -\arctan\left(\frac{v\sin\beta - l_{\mathrm{r}}\dot{\varphi}}{v\cos\beta}\right) \tag{12}$$

where δ is steering angle of front wheels and $\dot{\varphi}$ is the yaw rate of vehicle.

The traction force and lateral force of wheels in vehicle-fixed coordinate system, $F_{\mathrm{x,f}j}$, $F_{\mathrm{y,f}j}$, $F_{\mathrm{x,r}j}$ and $F_{\mathrm{y,r}j}$, can be presented by the transformation

$$F_{\mathrm{x,f}j} = F_{\mathrm{l,f}j}\cos\delta - F_{\mathrm{s,f}j}\sin\delta \tag{13}$$

$$F_{\mathrm{y,f}j} = F_{\mathrm{l,f}j}\sin\delta + F_{\mathrm{s,f}j}\cos\delta F_{\mathrm{x,r}j} = F_{\mathrm{l,r}j} \tag{14}$$

$$F_{\mathrm{y,r}j} = F_{\mathrm{s,r}j} \tag{15}$$

2.2 Vehicle dynamics

The vehicle dynamic of an NTV is the combination of a vehicle model and a bicycle model. In modelling an NTV system, the vehicle and its rider (and passenger) can be seen as a single mass model. The vehicle model of narrow tilting vehicle includes the velocity dynamic, side-slip angle dynamic, yaw dynamic and roll dynamic [30]. The geometry model of an NTV is shown as in **Figure 8**. The vehicle motion dynamics can be described by the vehicle velocity v and the vehicle side-slip

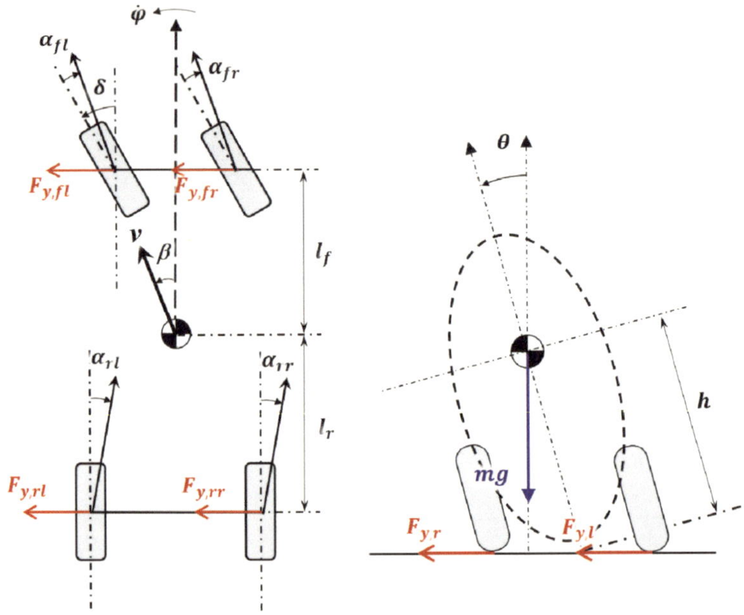

Figure 8.
Geometry of a narrow tilting vehicle.

angle β, which is defined as the angle between v and the vehicle longitudinal axis x. Their dynamics can be represented by

$$\dot{v} = \frac{1}{m}\left(\cos\beta\sum_{ij}F_{\mathrm{x},ij} + \sin\beta\sum_{ij}F_{\mathrm{y},ij} - F_{\mathrm{res}}\right) \tag{16}$$

$$\dot{\beta} = \frac{1}{mv}\left(\cos\beta\sum_{ij}F_{\mathrm{y},ij} - \sin\beta\sum_{ij}F_{\mathrm{x},ij}\right) - \dot{\varphi} \tag{17}$$

where F_{res} represents the force of driving resistance.

The vehicle acceleration in x- and y-axis a_{x} and a_{y} are presented by v, β, φ and their differentials as

$$a_{\mathrm{x}} = \dot{v}\cos\beta - v(\dot{\beta} + \dot{\varphi})\sin\beta \tag{18}$$

$$a_{\mathrm{y}} = \dot{v}\sin\beta + v(\dot{\beta} + \dot{\varphi})\cos\beta \tag{19}$$

The vehicle yaw motion in the second-order differential equation is represented as

$$\ddot{\varphi} = \frac{1}{I_z}\left[l_f\left(F_{\mathrm{y,fl}} + F_{\mathrm{y,fr}}\right) - l_r\left(F_{\mathrm{y,rl}} + F_{\mathrm{y,rr}}\right) + \frac{b_{\mathrm{f}}}{2}\left(F_{\mathrm{x,fr}} - F_{\mathrm{x,fl}}\right) + \frac{b_{\mathrm{r}}}{2}\left(F_{\mathrm{x,rr}} - F_{\mathrm{x,rl}}\right)\right] \tag{20}$$

where I_{z} is the inertia moment in z-axis.

As the NTV has no roll stiffness of suspension, the roll motion of NTV can be presented as

$$\ddot{\theta} = \frac{1}{I_x + mh^2\sin^2\theta}\left[mhg\sin\theta - h\cos\theta\sum F_{\mathrm{y},ij} - mh^2\dot{\theta}^2\sin\theta\cos\theta - C_{\mathrm{d}}\dot{\theta}\right] \tag{21}$$

where θ and $\dot{\theta}$ are the vehicle roll angle and roll rate, I_x is the vehicle roll moment of inertia, and C_d is the roll damping ratio of the suspension.

2.3 Simplified single-track vehicle model

The nonlinear equations of the four-wheel model provided in previous section are accurate and detailed in matching the real vehicle response. In controller design and performance analysis, a simplified single-track model has been delivered from the nonlinear equations (1)–(21). Assuming that the steer angle, side slip angle and roll angle approach zero at normal states, their sinusoidal value can be approximated to their own value using the small-angle approximation for simplification. And assume that the COG is at the middle of the track and the difference between COG to the front and rear axles is zero, which gives $l_f = l_r$. Simplify the foundation torque of rear left and rear right wheels as T_r and add the torque differential value of rear wheels as a new input ΔT_r to the system. Then the vehicle model can be simplified to a function of system state x and control input u as

$$\dot{x} = f(x) + g(x) \cdot u \tag{22}$$

where

$$x = \left[v \ \beta \ \dot{\varphi} \ \theta \ \dot{\theta}\right]^T, \quad u = \left[\delta T_r \Delta T_r\right]^T \tag{23}$$

$$f(x) = \begin{bmatrix} -\frac{2C_\gamma}{m}\beta^2 + \frac{2\lambda_\gamma}{m}\beta\theta \\ -\frac{2C_\gamma}{m}\frac{\beta}{v} + \frac{2\lambda_\gamma}{m}\frac{\theta}{v} - \dot{\varphi} \\ -\frac{C_\gamma l^2}{2I_z}\frac{\dot{\varphi}}{v} \\ \dot{\theta} \\ \frac{mgh - 2\lambda_\gamma h}{I_x}\theta - \frac{C_d}{I_x}\dot{\theta} + \frac{2C_\gamma h}{I_x}\beta - \frac{mh^2}{I_x}\dot{\theta}^2\theta \end{bmatrix} \tag{24}$$

$$g(x) = \begin{bmatrix} \frac{C_\gamma}{m}\beta & \frac{C_\gamma}{mv} & \frac{C_\gamma l\beta}{2I_z} & 0 & -\frac{C_\gamma h}{I_x} \\ \frac{1}{mR_r} & -\frac{\beta}{mv} & 0 & 0 & 0 \\ 0 & 0 & \frac{b_r}{I_z} & 0 & 0 \end{bmatrix}^T \tag{25}$$

including the linearised tyre lateral behaviour as equivalent cornering stiffness coefficient C_γ and camber stiffness coefficient λ_γ.

The system will finally converge to its steady state with a given trajectory. When the vehicle is turning in a circle with radius of R, the system steady state value can be approximately calculated as

$$\begin{cases} \beta_0 = l/2R \\ \dot{\varphi}_0 = v/R \\ \theta_0 = v^2/gR \end{cases} \tag{26}$$

2.4 Virtual rider model

The virtual rider model is produced to simulate the reaction of a rider of NTV with two objective, one is to maintain the stability of vehicle to reduce the risk of falling down and the other is to follow the path of the target route [9, 37]. For experienced rider, the two control objectives can be achieved together to perform the optimised operation. However, the virtual rider model needs to be developed assuming the rider has no special skills and experience in operating an NTV [2, 16, 38]. A solution is to apply two control algorithms independently, one aims to maintain the roll angle of the vehicle and the other aims to track the path. Each control algorithm has only one control objective and will not communicate with each other to simulate the behaviour of a new rider to ride an NTV. Different with riding a bicycle that the rider can shift its own body to help lean the bicycle, the weight of NTV is much higher than the weight of a rider. Thus, the shift of rider's position is not considered in the virtual rider and only the steer angle and traction torque are controlled by the virtual rider.

The steer angle control will result in not only the roll stability but also the lateral dynamics. Because of this, the control algorithm of steer angle can be easily developed as the sum of two simple controllers, one tracking the roll angle θ and the other tracking the yaw rate $\dot{\varphi}$. In roll stability control, a proportional derivative (PD) control is applied to track the error of roll angle [3] as

$$\delta_1 = \left(k_{p2} + sk_{d2}\right)\left(\theta_{ref} - \theta\right) \tag{27}$$

In the lateral trajectory tracking, the rider applies steering input to track the target yaw rates obtained from path. It is assumed that the steer angle is proportional with the path to be followed [3]. As the required response speed of lateral trajectory tracking is slower than that of roll stability control, a pseudo-derivative feedback (PDF) control is applied in the yaw rate tracking. Comparing with the traditional PI (D) control, the PDF control can reduce the effect of derivative feedforward action to avoid the transient impact to roll stability [39]. The lateral control is designed as

$$\delta_2 = \frac{k_{i1}}{s}\left(\dot{\varphi}_{ref} - \dot{\varphi}\right) - k_{p1}\dot{\varphi} \tag{28}$$

The final steer angle control input can be calculated with combining the two outputs together as

$$\delta = \delta_1 + \delta_2 \tag{29}$$

Apart from the steering control to maintain the roll stability and follow the path, the vehicle speed needs to be controlled by virtual rider via throttle to generate equivalent traction torque to the vehicle. The vehicle velocity control is implemented via a PI controller as

$$T_r = \left(k_{p3} + \frac{k_{i3}}{s}\right)\left(V_{ref} - v\right) \tag{30}$$

3. Design of drive assistance system

This section includes two designs. The first design is a nonlinear tilting controller which compensates the nonlinear behaviour of DTC under different vehicle velocity and operating state. The second design is two torque vectoring controllers to assist the rider to maintain the roll dynamics of NTV in cornering based on steer angle and tilting compensator, respectively.

3.1 Nonlinear tilting controller design

In the nonlinear controller design, the first step is to linearise the relation between system input and output, which is called input-output linearisation. The roll angle equation can be rewritten as

$$\ddot{\theta} = L_f(x) + B(\theta)M_t \tag{31}$$

where

$$\begin{aligned} L_f(x) = & \frac{1}{I_x + mh^2 \sin^2\theta} \Big[mhg \sin\theta - mh^2\dot{\theta}^2 \sin\theta\cos\theta \\ & -h\cos\theta\Big(2C_f\delta - 2C_f \tan^{-1}\Big(\frac{l_f\dot{\varphi} + V_y}{V_x}\Big) \\ & +2\lambda_f\theta + 2C_r \tan^{-1}\Big(\frac{-l_r\dot{\varphi} + V_y}{V_x}\Big) + 2\lambda_r\theta\Big)\Big] \end{aligned} \tag{32}$$

$$B(\theta) = \frac{1}{I_x + mh^2 \sin^2\theta} \tag{33}$$

As $B(x) \neq 0$ if $I_x > 0$, the $B(x)$ is known as non-singular for nominal operating points and $B(x)^{-1}$ is achievable. The required tilting motion as control input can be designed as

$$M_t = B^{-1}(-L_f(x) + u_t) \tag{34}$$

where u_t is designed as the control input to the linearised system

$$u_t = k_1(\theta^* - \theta) - k_2\dot{\theta} \tag{35}$$

and θ^* is the ideal roll angle calculated as a function of steering angle and vehicle speed as

$$\theta^* = \tan^{-1}\left(\frac{V^2\delta}{(l_f + l_r)g}\right) \tag{36}$$

Choose $B_0 = B(\theta)\|_{\theta=0}$ as the nominal control gain at rated value. The difference between $B(x)$ and B_0 will be seen as disturbance in the lumped perturbation. Define the perturbation terms as

$$\ddot{\theta} = \Psi(x) + B_0M_t \tag{37}$$

where

$$\begin{aligned} \Psi(x) &= L_f(x) + (B(x) - B_0)M_t \\ B_0 &= \frac{1}{I_x} \end{aligned} \tag{38}$$

Assume the lumped perturbation will not change quicker than one time cycle. Calculate the perturbation term as

$$\hat{\Psi}(t) \approx \Psi(t-1) = \ddot{\theta}(t-1) - B_0 M_t(t-1) \tag{39}$$

The final control input is

$$M_t = B_0^{-1}\left[-\hat{\Psi} + k_1 \tan^{-1}\left(\frac{V_x^2 \delta / g}{l_f + l_r}\right) - k_1\theta - k_2\dot{\theta}\right] \tag{40}$$

Figure 9 shows the block diagram of the nonlinear control for tilting mechanism of NTV.

3.2 Torque vectoring controller design

3.2.1 Steering angle based torque vectoring (SATV)

The easiest design to compensate the counter steering behaviour when turning a vehicle is to set the vectoring torque of rear wheels proportionally to the derivative of steer angle as

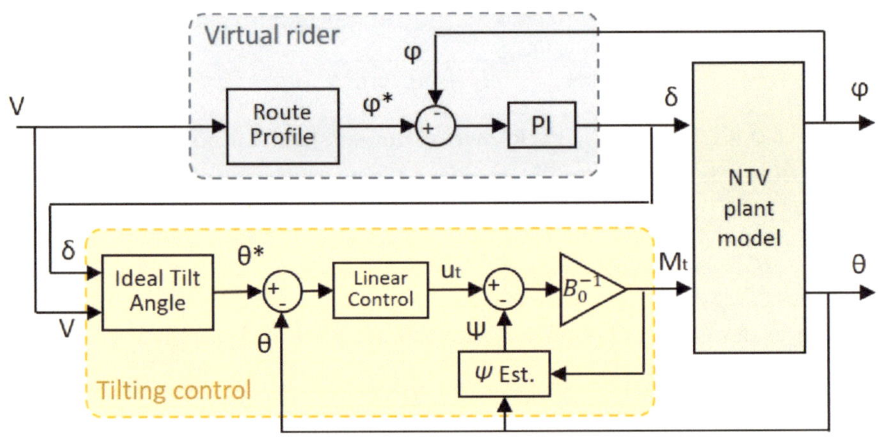

Figure 9.
Control block diagram of the tilting control for the tilting mechanism of NTV.

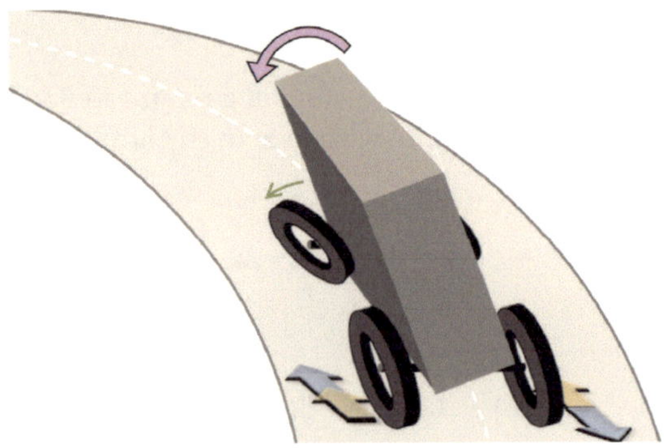

Figure 10.
Vectoring torque assists the rider in balancing the NTV during a turn.

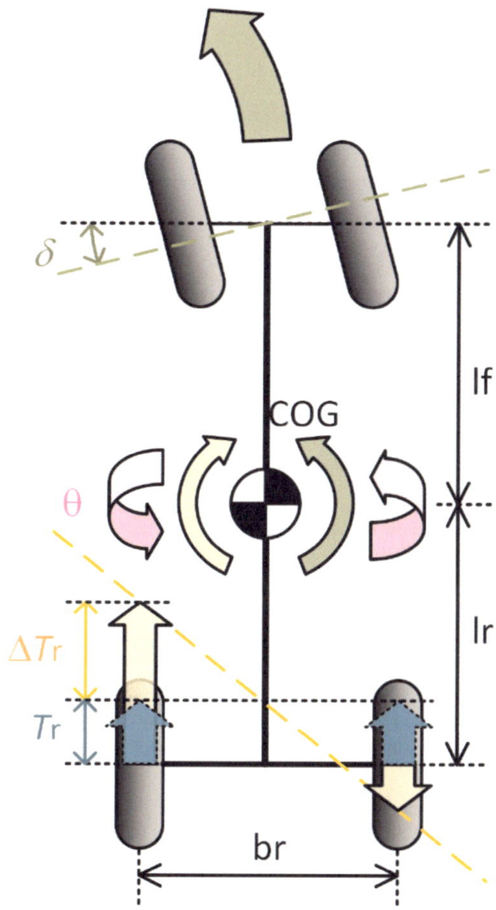

Figure 11.
The diagram of torque vectoring for narrow tilting vehicle.

$$\Delta T_{\mathrm{r}} = K\dot{\delta} \tag{41}$$

where K is the control gain designed to get expected performance. This parameter is chosen based on the bandwidth of the torque controller that its response speed have to be several times faster than yaw moment response speed and slower than torque response speed.

When the rider turning the vehicle, the torque controller activated the vectoring torque to yaw the vehicle in the opposite direction to lean the vehicle to the target position (**Figures 10** and **11**). With the drive assistance of vectoring torque, the rider is not required to act counter-steering to lean the vehicle manually.

3.2.2 Tilting compensator based torque vectoring (TCTV)

In steady-state that the vehicle keeps a unchanged states, $\ddot{\varphi}$ reaches zero and $\dot{\varphi}$ reaches its reference value. Then the steady-state steer angle can be obtained from the yaw dynamics in (22) as

$$\delta_{\mathrm{ss}} = \frac{l}{v}\dot{\varphi} - \frac{2b_{\mathrm{r}}}{C_{\gamma}l}\Delta T_{\mathrm{r}} \tag{42}$$

Substitute (42) into the roll dynamic equation in (22) to obtain:

$$\ddot{\theta} = \frac{1}{I_x}\left[(mgh - 2\lambda_\gamma h)\theta - C_d\dot{\theta} + 2C_\gamma h\beta - mh^2\dot{\theta}^2\theta \right. \\ \left. -C_\gamma h\left(\frac{l}{v}\dot{\varphi} - \frac{2b_r}{C_\gamma l}\Delta T_r\right)\right] \tag{43}$$

Assume $\dot{\theta}$ and $\ddot{\theta}$ are zero in steady state, one can obtain the equation below:

$$\ddot{\theta} = \frac{h}{I_x}\left[(mg - 2\lambda_\gamma)\theta + 2C_\gamma\beta - \frac{C_\gamma l}{v}\dot{\varphi} + \frac{2b_r}{l}\Delta T_r\right] = \sigma \tag{44}$$

If design the control signal as

$$\sigma = \frac{h}{I_x}\frac{2b_r}{l}K\dot{\delta} = K'\dot{\delta} \tag{45}$$

the vectoring torque to improve roll stability can be obtained as

$$\Delta T_r = K\dot{\delta} + \Psi \tag{46}$$

where

$$\Psi = \frac{l}{2b_r}\left[C_\gamma\delta - (mg - 2\lambda_\gamma)\theta - 2C_\gamma\beta\right] \tag{47}$$

Comparing (46) with (41), the additional component Ψ is defined as the tilting compensator (TC) to compensate the nonlinear impacts that could reduce the roll stability. The TC based torque vectoring (TCTV) method can manage the vectoring torque to reduce the counter steering during a turn. The block diagram of the TV based drive assistance system is shown in **Figure 12**.

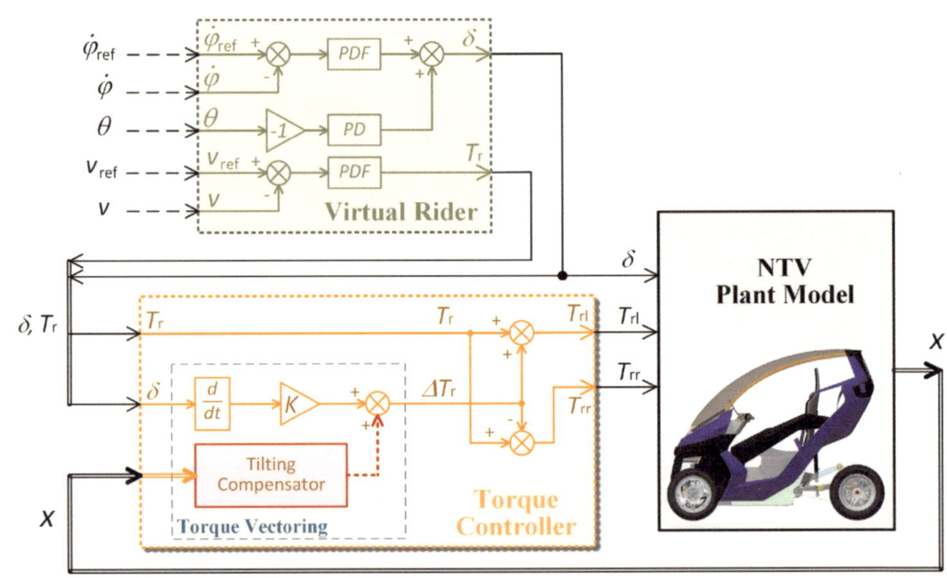

Figure 12.
The control block diagram of torque vectoring.

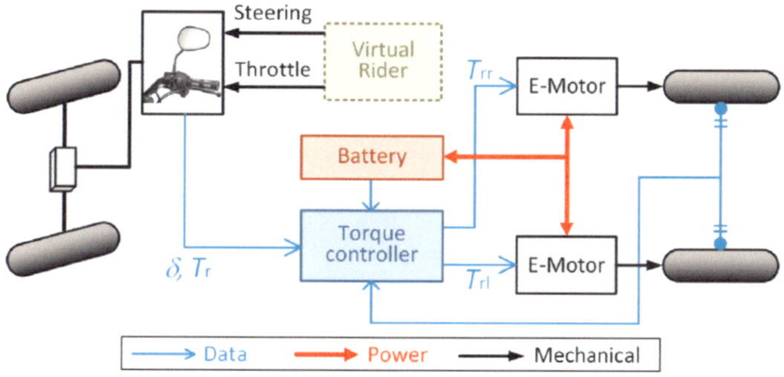

Figure 13.
The data communication and power flow of torque vectoring in narrow tilting vehicle.

3.2.3 Torque management

For electric vehicles, the main source of vehicle propulsion is from the batteries. To protect the battery and electric motors from overload, the output power will be limited to manage the generated torque in the controller. The available torque applied to the motor can be described as

$$T_{\text{avi}} = \min\left(T_{\text{m,rated}}, \frac{\min\left(P_{\text{m,rated}}, P_{\text{b,avi}}\right)}{\omega_m}\right) \tag{48}$$

where $T_{\text{m,rated}}$ and $P_{\text{m,rated}}$ is the rated torque and power of wheel motor; $P_{\text{b,avi}}$ is the maximum available output power from battery based on its current capacity. The final torque output can be managed as

$$T'_{\text{r}} = \min\left(T_{\text{r}}, T_{\text{avi}}\right) \tag{49}$$

$$\Delta T'_{\text{r}} = \min\left[\Delta T_{\text{r}}, \left(T_{\text{avi}} - T'_{\text{r}}\right)\right] \tag{50}$$

Then the final torque applied on the left and right rear wheels can be represented as

$$\begin{cases} T_{\text{rl}} = T'_{\text{r}} + \Delta T'_{\text{r}} \\ T_{\text{rr}} = T'_{\text{r}} - \Delta T'_{\text{r}} \end{cases} \tag{51}$$

The torque drive system of NTV is shown in **Figure 13**, where the data flow, electric power flow, and mechanical drive are given with blue, red and black arrows, respectively.

4. Simulation results

The following section verifies the designed controllers in assisting the rider to maintain the roll stability of the vehicle during a turn. The parameters of NTV are chosen from [1] and given in **Table 1**. The verification is based on the simulation in Matlab.

4.1 Result of nonlinear tilting control approach

4.1.1 Constant speed tilting test

The controllers used in simulation are with the parameters given in **Table 2**. The simulation validations of nonlinear tilting controller are carried out with tracking the

Description	Symbol	Value	Unit
Total vehicle mass	m	200.0	kg
Height of vehicle COG	h	0.5	m
Distance from COG to front axle	l_f	0.7	m
Distance from COG to rear axle	l_r	0.9	m
Length of track of front axle	b_f	0.5	m
Length of track of rear axle	b_r	0.7	m
Vehicle roll moment inertia	I_x	18	kg·m^2
Vehicle yaw moment inertia	I_z	80	kg·m^2
Front/Rear wheel radius	$R_{\text{fj/rj}}$	0.5	m
Front/Rear wheel rotational inertia	$J_{\text{fj/rj}}$	0.2	kg·m^2
Front cornering stiffness	C_f	3500	N/rad
Rear cornering stiffness	C_r	5480	N/rad
Front camber stiffness	λ_f	1000	N/rad
Rear camber stiffness	λ_r	2000	N/rad

Table 1.
System parameters of NTV.

Description	Symbol and value
Virtual rider	$K_\text{p} = 0.1, \quad K_\text{i} = 0.1$
Linear controller	$k_1 = 300, \quad k_2 = 400$
Gain-scheduling controller	$\begin{cases} k_1 = 300, \quad k_2 = 400 \quad (LS: v \le 18\,km/h) \\ k_1 = 500, \quad k_2 = 1000 \quad (MS: 18 < v \le 30\,km/h) \\ k_1 = 1500, \quad k_2 = 3000 \quad (HS: v > 30\,km/h) \end{cases}$
Nonlinear controller	$k_1 = 300, \quad k_2 = 400, \quad B_0 = 0.0556$

Table 2.
Control parameters.

route in a shape of $'\infty'$. The speed reference gives a constant and the yaw rate reference is in a square wave. The virtual rider maintains the vehicle yaw rate to track the predefined route path that is independent with the tilting controller. The route path tracking performance is shown in **Figure 14**, where the desired path is shown as the black line and the vehicle path is shown as the dashed red line. The vehicle path track is due to the yaw rate control of a virtual rider and the stability is controlled by the tilting mechanism. The entire control performance is good that the vehicle tracks the path well.

The vehicle yaw rate and roll angle performance at the constant speed of 20 km/h are shown as the simulation result in **Figure 15**. The yaw rate tracks the step changed reference by the virtual rider and the vehicle roll angle is controlled by the nonlinear tilting controller. The result in **Figure 15** compares between the control performance (blue line) and their reference (red dashed line). The result of roll angle tracking indicates that the controller is able to keep the vehicle stable during a turn.

4.1.2 Increasing and decreasing speed tilting test

In the increasing and decreasing speed tilting test, the same route path is employed for the validation with the varying vehicle velocity from low speed to

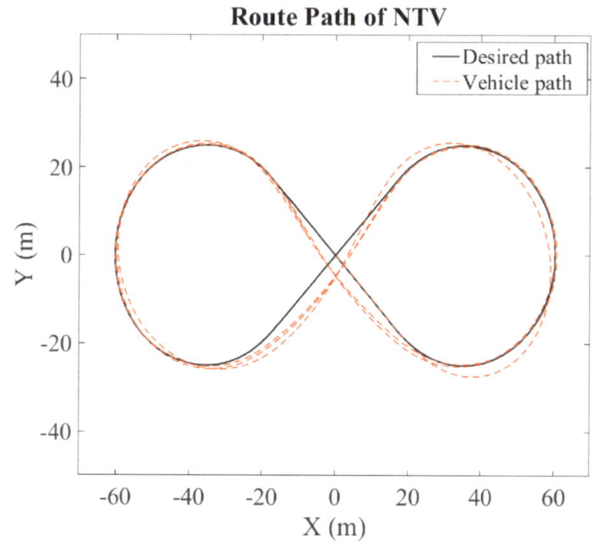

Figure 14.
Simulation result of vehicle route path tracking performance.

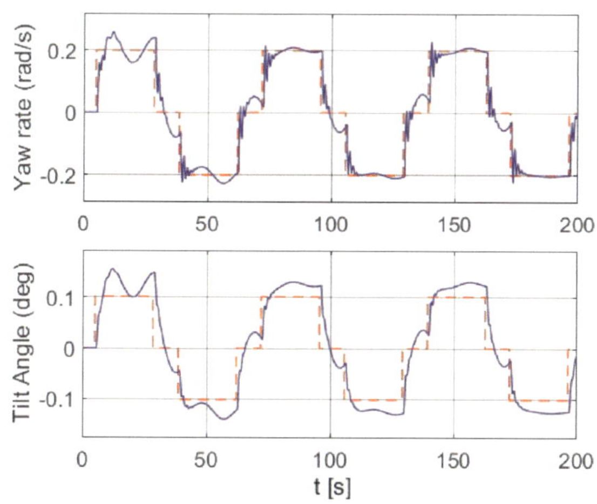

Figure 15.
Simulation result of vehicle yaw rate and roll angle at constant speed of 20 km/h.

high speed. With the change of vehicle speed, the reference yaw rate and roll angle are amended depending on the velocity. Its yawing and tilting dynamics have nonlinear behaviour under different speed. Three types of controller have been compared, the linear PI controller, the gain-scheduling PI controller, and the nonlinearity compensation based tilting controller.

The linear controller normally cannot perform well in the whole range of speed, especially at high vehicle speed. That is because the vehicle model includes high-order nonlinearities and the linear controller is designed based on the linearised model at a particular operating speed. Thus, the control performance can be guaranteed only at the target speed. The nonlinear effect, such as the variation of speed, will cause the controller difficult to maintain the optimised performance in the whole speed range.

To reduce the nonlinear effect at different operating speed, the gain-scheduling (GS) controller is to separate the whole speed region into small regions and the linear control law is applied in each region. When the vehicle velocity changed from low speed to high speed, the controller parameters switch among the predesigned regions to provide the desired control performance in the whole range of speed. The scheduled control parameters in three regions, low speed (LS), medium speed (MS) and high speed (HS), are given in **Table 2**. However, the switching between controllers reduces the control performance and the predefined controllers are not robust to the uncertainties in practice, including both the variation of vehicle parameters, such as the weight and COG of vehicle caused by the change of rider and passengers, and the environmental variation, such as the friction coefficient of road surface and impact of wind disturbance. As the simulation aims to verify the control algorithm under ideal condition, the impact of uncertainties is not presented in the simulation study.

The nonlinear controller is designed to cover the whole operating region from low vehicle velocity to high vehicle velocity. The structure of the nonlinear controller is the combination of a normal linear controller and a compensation block of estimated nonlinearity, as in **Figure 9**. The nonlinearity under different speed was compensated by the estimated nonlinear behaviour. The nonlinear approach improved the performance of tilting controller in the whole operating region.

In the chosen of controller parameters, the lower control gain can cause larger tracking error, while the higher control gain can cause more actuator usage and overshoots and can reduce the stability. The parameters of linear controller are chosen to minimise the roll angle tracking error in the whole operating range and, simultaneously, reduce the risk of causing unstable. To ensure a fair comparison, the GS control gain in each region is chosen by the method the same as that of the linear controller. And the nonlinear controller uses the same control gain as the linear controller additionally combined with the nonlinearity compensation block. The simulation results of vehicle yaw rate and roll angle controlled by the three controllers under the speed from 5 to 45 km/h are shown in **Figures 16–18**, respectively.

To verify the improvement of control performance numerically, the comparison among linear, gain-scheduling and nonlinear controllers is relied on the integral absolute error (IAE) of yaw rate and roll angle, as shown in **Figure 19**. All the three controllers are employed to control the same vehicle system with the same virtual rider model. The roll angle IAE of the nonlinear controller is 46% less than that of the GS controller and 75% less than that of the linear controller. In addition, the better performance the tilting controller acted on roll angle tracking, the easier the rider can maintain the vehicle in tracking the yaw rate.

With the same virtual rider, the performance of nonlinear controller has 24 and 9% less IAE in yaw rate tracking than that of the linear controller and GS controller, respectively. The results verified that the nonlinear tilting controller performs better in both maintaining the roll stability and tracking the lateral trajectory.

4.2 Result of torque vectoring approach

In the simulation validation, the TV controllers are applied to control the roll stability of NTV in tracking the route of a step yaw rate in two case studies. The first case is that the vehicle driven into a turn at a constant speed and the second case is that the vehicle accelerating during a turn. Both tests use the same rider model and vehicle plant model in the comparison among traditional controllers and proposed SATV and TCTV controllers. The parameter settings of the virtual rider model and torque controller are given in **Table 3**. The performance validates the effectiveness of the designed controller on counter steering reduction and the stability improvements of the designed controllers.

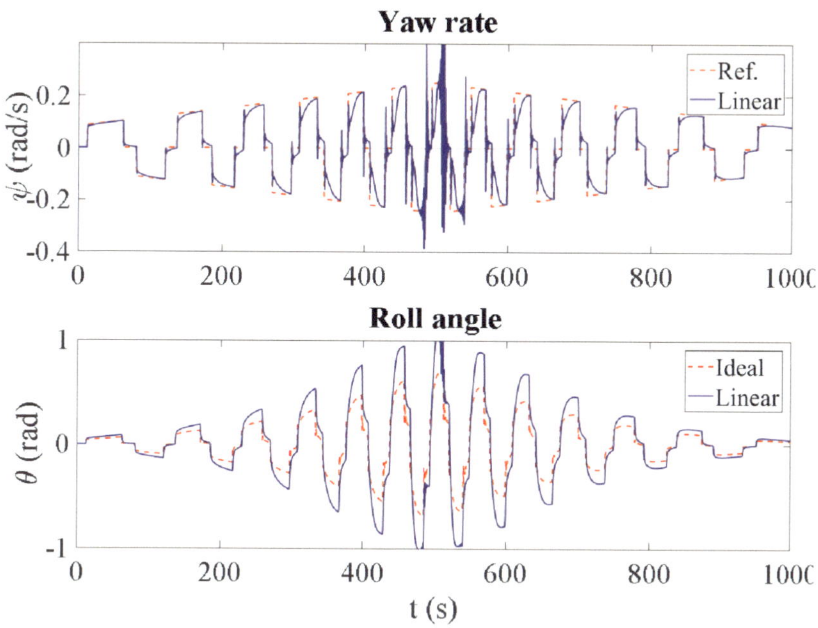

Figure 16.
Simulation result of yaw rate and roll angle with linear controller.

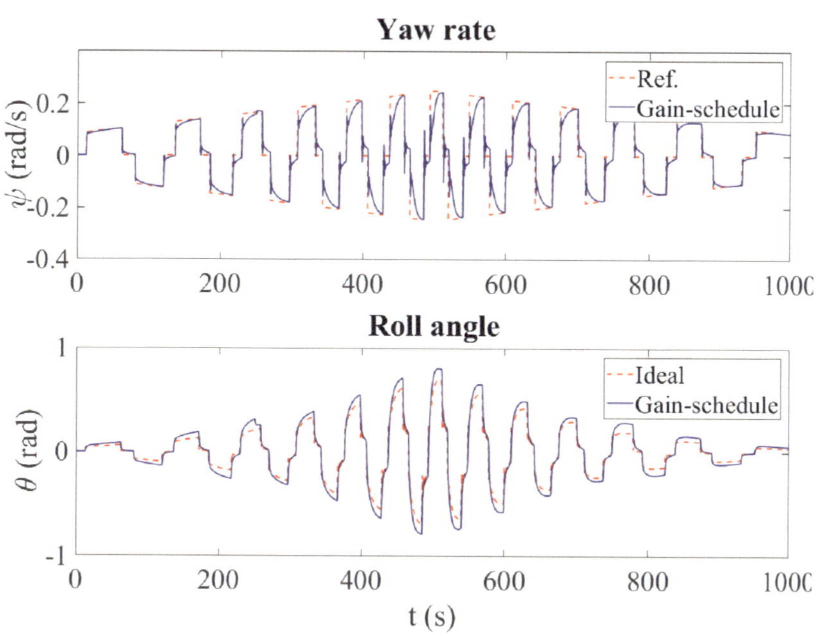

Figure 17.
Simulation result of yaw rate and roll angle with gain-scheduling controller.

4.2.1 Left turn under constant speed

The case study simulates the dynamic response of an NTV driving into a turn. The vehicle is driven straight with a constant speed of 5 m/s at first. Then the rider starts to turn the vehicle to track the path in a left turn with the radius of 15 m, as

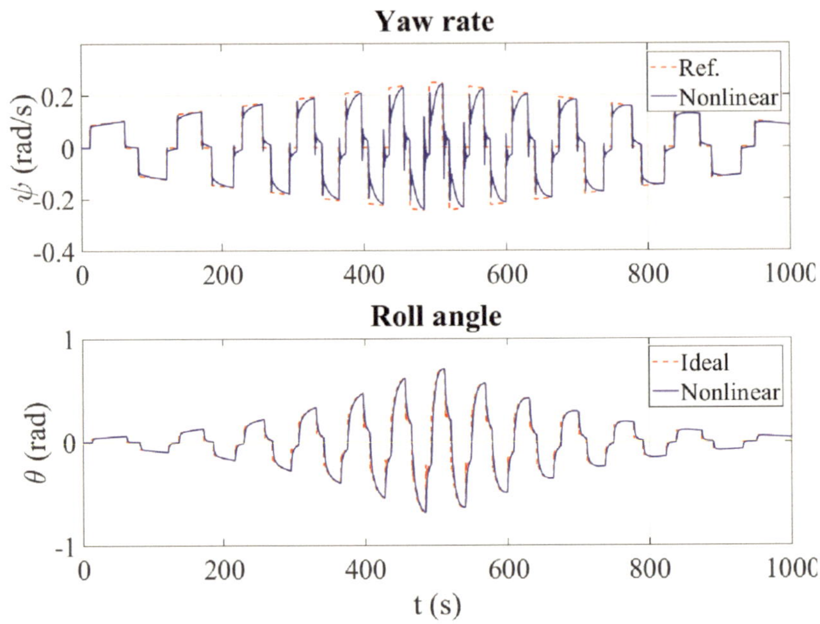

Figure 18.
Simulation result of yaw rate and roll angle with nonlinear controller.

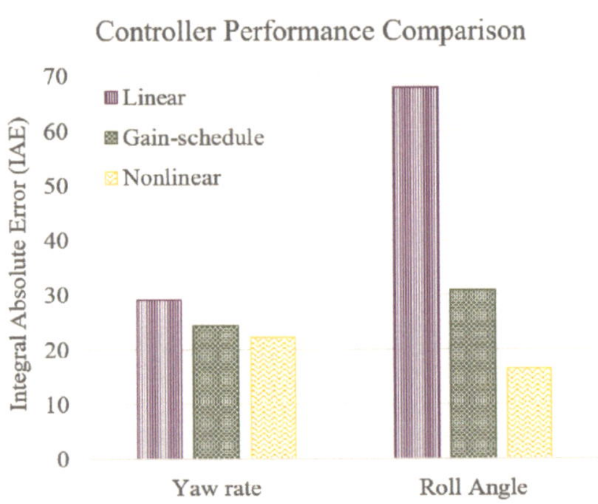

Figure 19.
Controller performance comparison among their IAE of yaw rate and roll angle.

Description	Symbol and value	
Virtual rider	k_{p1} = 0.3	k_{i1} = 0.2
	k_{p2} = 1	k_{d2} = 5
	k_{p3} = 1	k_{i3} = 0.4
Torque controller	K = 50	
	$T_{m,rated}$ = 50 Nm	
	$P_{m,rated}$ = 1500 W	

Table 3.
Controller parameter settings.

shown in **Figure 20**. The desired command to the virtual rider is a step changed yaw rate. The step change of yaw rate reference actually acts as a sudden disturbance to the torque controller to verify its transient response. In conventional method, the rider should counter-steer the front wheels to lean the vehicle into an opposite direction until the roll angle reaches the desired value to maintain its roll stability. With the assistance of torque vectoring, the requirements of counter-steer from rider will be reduced.

Figure 21 shows the dynamic response of the steering angle, vectoring torque, vehicle side-slip angle, yaw rate, lateral acceleration and roll rate. The comparisons are among the control inputs as well as the vehicle states performance under the control of the different controllers. Comparing the results, the requirements of counter-steering from rider have been eliminated in both the SATV and TCTV based control methods. Comparing between the two proposed TV methods, the TCTV has less oscillation than that of the SATV due to the tilting dynamics been compensated.

The steady state value of the target yaw rate is 19 deg/s and that of the side-slip angle is 2.9 deg, which can be calculated from (26). The yaw rate and lateral acceleration of the TCTV based torque control have less oscillation comparing with the other three methods. The steering angle is the main contributor to the performance of side-slip angle. This makes them to have the same response as shown in the results. Among the performances of all the controllers, the roll rate of the TCTV based torque control performs the best with the least peak tracking error and less oscillation.

The tracking error of all states is shown in **Figure 22** for a clearer comparison. In the comparison of tracking performance, the proposed controller performs better in transient response with less maximum tracking error and oscillation rate. The tracking error compensation speed is about 4 seconds to approach zero under disturbance. It verifies that both the transient stability and steady-state stability of roll dynamics can be maintained as desired with using the proposed controller.

The quantity comparison of maximum tracking error and integral absolute error (IAE) of each state is summarised and given in **Table 4**. With the usage of proposed TV controller, both the maximum tracking error and oscillation rate has been reduced comparing with traditional control method and normal TV approach. The TCTV eliminated the counter-steering process to ease the rider in operating the NTV and the maximum error of steering control is reduced about 74% comparing

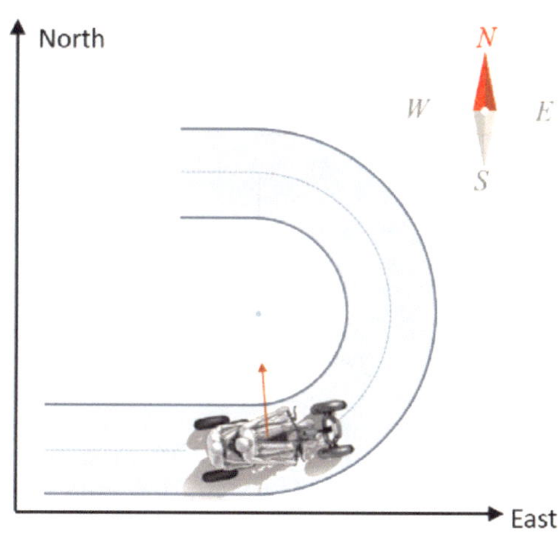

Figure 20.
Path of vehicle with left turn in simulation.

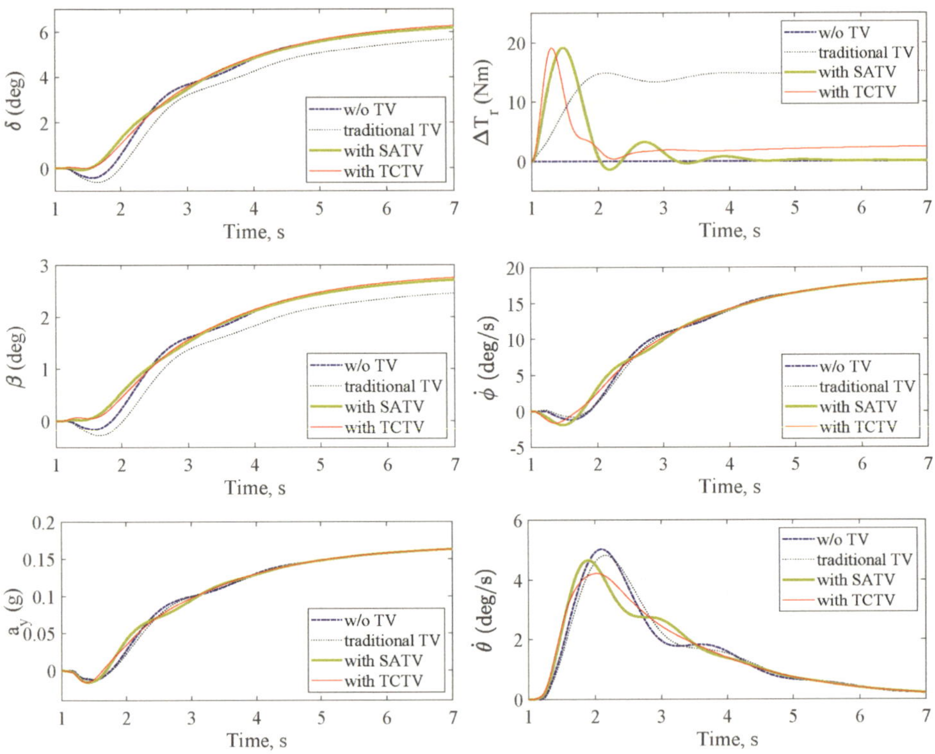

Figure 21.
Simulation result of case 1—left turn under constant speed.

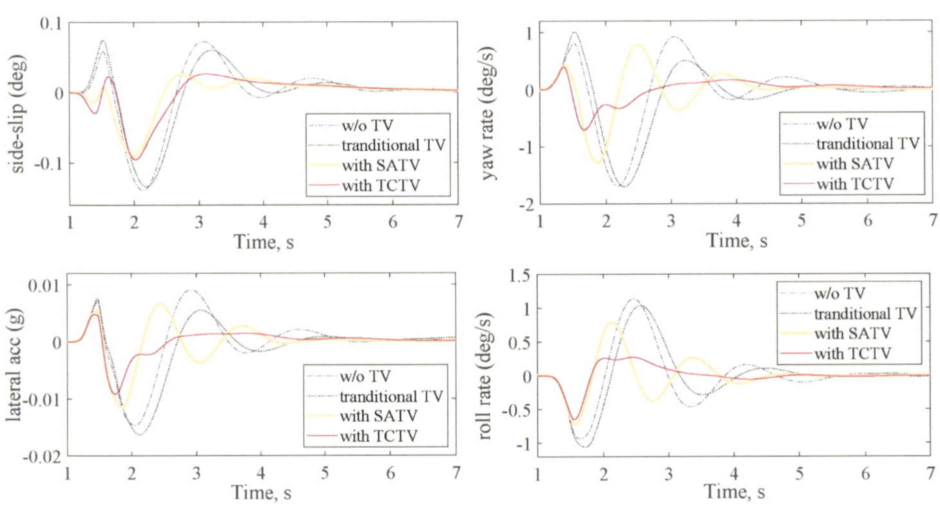

Figure 22.
States tracking error comparison of case 1.

with the result of conventional method. With the TCTV controller applied to the NTV, the side-slip angle, yaw rate, lateral acceleration, and roll rate have been improved with 35, 58, 36, 28% less maximum tracking error, respectively. To make the comparison more obvious, the indices of the maximum error, IAE and oscillation rate of all the control methods are shown in bar charts as in **Figure 23**.

Indices	Variables	w/o TV	Traditional TV	With SATV	With TCTV
Maximum track error	Counter-steer agl (deg)	0.553	0.311	0.107	0.006
	Side-slip agl (deg)	0.1442	0.138	0.0943	0.101
	Yaw rate (deg/s)	1.763	1.82	1.307	0.719
	Lateral acc (×0.01 m/s)	1.51	1.711	1.19	0.933
	Roll rate (deg/s)	1.166	1.086	0.803	0.653
Integral absolute error	Side-slip agl (deg·s)	0.297	0.290	0.160	0.185
	Yaw rate (deg)	3.66	3.43	2.07	1.24
	Lateral acc (×0.01 m/s)	3.136	3.217	1.861	1.199
	Roll rate (deg)	2.586	2.426	1.52	0.832
Oscillation rate	Counter-steer agl (%)	0.6093	0.6343	0.5442	0.5292
	Side-slip agl (%)	0.6807	0.7161	0.6082	0.5937
	Yaw rate (%)	0.4694	0.2709	0.2506	0.1829
	Lateral acc (%)	0.3912	0.205	0.1896	0.1217
	Roll rate (%)	3.691	2.123	1.947	1.546

Table 4.
Performance indices comparison among different controllers in turning at a constant speed.

4.2.2 Speed acceleration during a turn

In addition to the constant speed turn, the speed acceleration/deceleration will cause instability of the vehicle roll dynamics as well. The second case simulates the operating of NTV under the condition of speed accelerating during a turn. The initial state is the NTV driving at a constant speed of 5 m/s and turning left with a yaw rate of 5.8 deg/s in steady state. Then the rider increase the propulsion torque to accelerate the vehicle to test the response of the torque controller as well as the performance of the vehicle.

The dynamic response of an NTV is shown in **Figure 24** with two inputs and four system states. And the tracking error of all states is shown in **Figure 25**. The result in this case is similar to the previous one that both the SATV and TCTV reduced the requirements of counter-steering from rider and improves the roll stability with less tracking error. In the yaw rate and roll rate comparison, the TCTV performs the best with the least peak error and fastest response time. The numerical results and bar chart comparison of the maximum tracking error, IAE and oscillation rate are given in **Table 5** and **Figure 26**. The TCTV method reduced the maximum tracking error in steer angle, side-slip, yaw rate, lateral acceleration and roll rate with 35%, 44%, 59%, 73% and 55% less value, respectively.

The cases aim to verify the control performance of developed torque controller in sudden turning and speed acceleration during a turn. These two cases verifies the vehicle stability with the proposed torque controller under sudden disturbance and time-varying disturbance, respectively. In both cases, the counter steering requirements can be fully eliminated and the maximum tracking error and oscillation rate of state tracking performance can be reduced for the average of 1/3 with using the proposed TV approach. Comparing the two cases, the improvement is more obvious

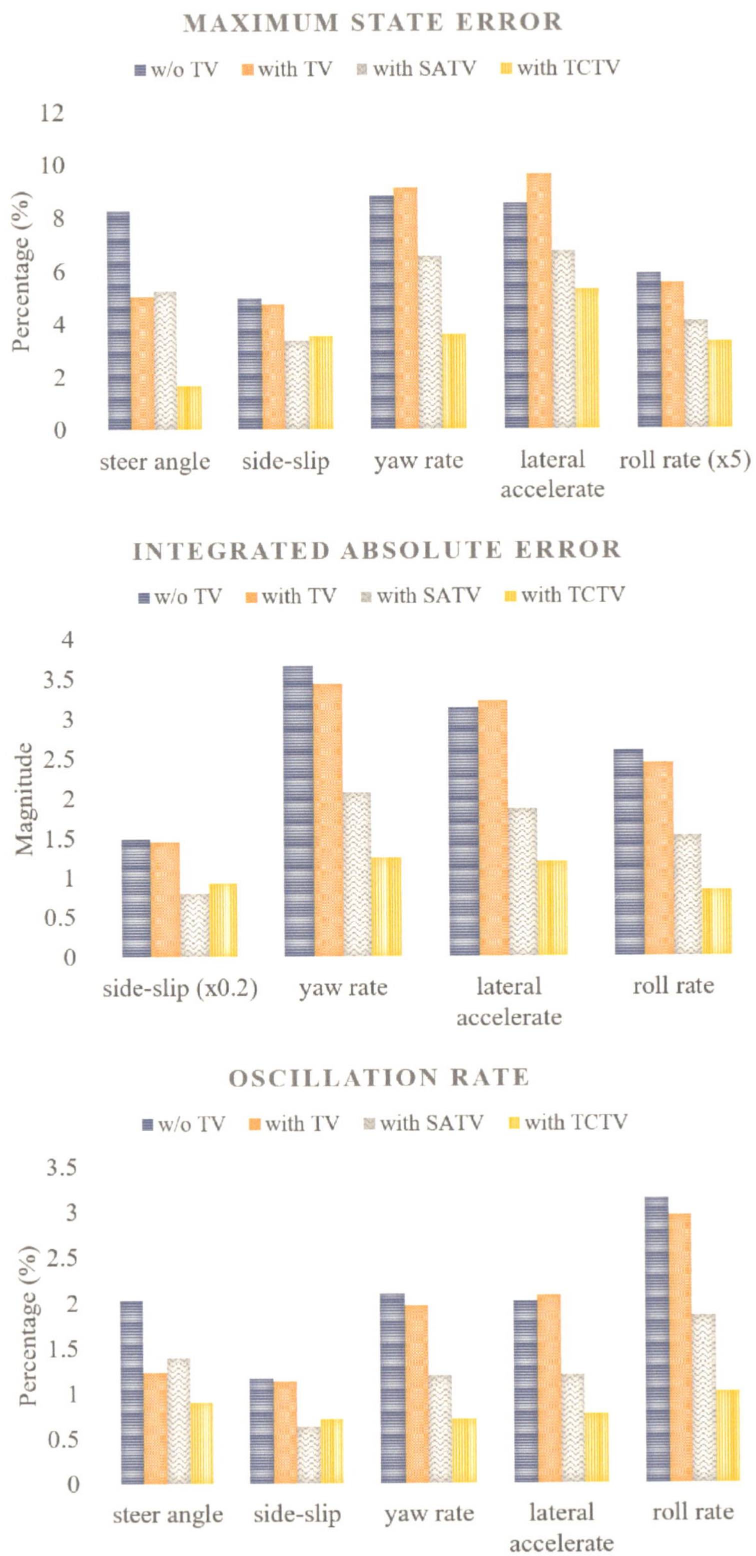

Figure 23.
The performance indices comparison among different torque controllers in the turning.

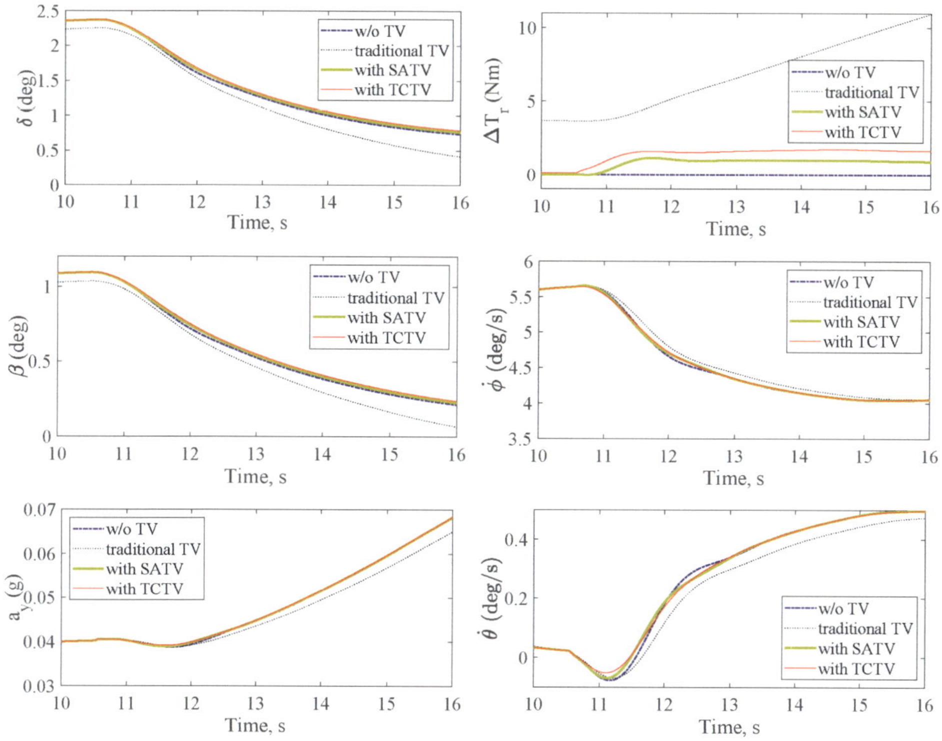

Figure 24.
Simulation result of case 2—acceleration during a turn.

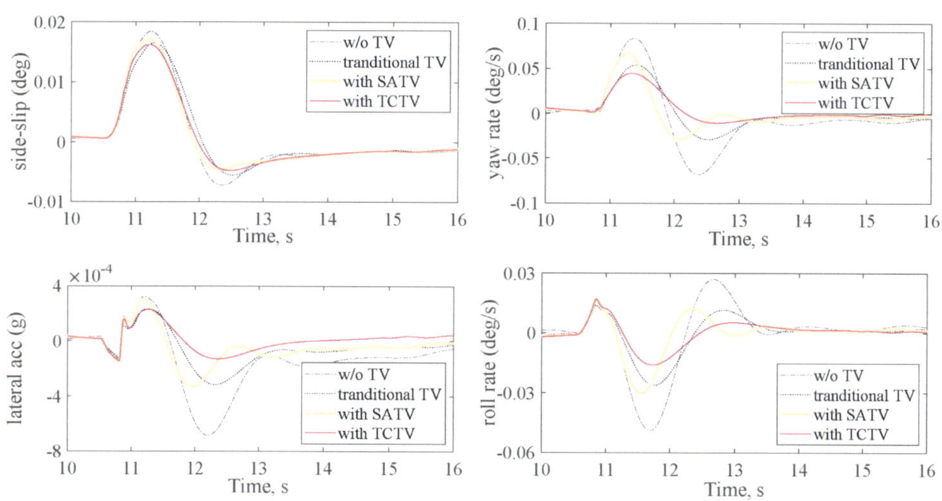

Figure 25.
States tracking error comparison of case 2.

in speed acceleration during a turn as it is more suitable for the time-varying disturbance on roll stability. Therefore, the NTV equipped with the drive assistance system will be easy to be ridden by any types of riders from new to experienced ones.

Indices	Variables	w/o TV	Traditional TV	With SATV	With TCTV
Maximum track error	Counter-steer agl (deg)	0.053	0.013	0.0027	0
	Side-slip agl (deg)	0.0184	0.0165	0.0173	0.0162
	Yaw rate (deg/s)	0.0834	0.0539	0.0673	0.0447
	Lateral acc (×0.01 g)	0.0687	0.0316	0.0332	0.0229
	Roll rate (deg/s)	0.049	0.0261	0.0302	0.0169
Integral absolute error	Side-slip agl (deg·s)	0.0266	0.0246	0.0242	0.0239
	Yaw rate (deg)	0.137	0.0793	0.0733	0.0535
	Lateral acc (×0.01 m/s)	0.118	0.0619	0.0572	0.0367
	Roll rate (deg)	0.0628	0.0332	0.0263	0.0361
Oscillation rate	Counter-steer agl (%)	0.6093	0.6343	0.5442	0.5292
	Side-slip agl (%)	0.6807	0.7161	0.6082	0.5937
	Yaw rate (%)	0.4694	0.2709	0.2506	0.1829
	Lateral acc (%)	0.3912	0.205	0.1896	0.1217
	Roll rate (%)	3.691	2.123	1.947	1.546

Table 5.
Performance indices comparison among different controllers in accelerating in a turn.

However, one limitation of this work is that it only simulated the driving assistant system in NTV under ideal condition, more uncertainty impact as well as the experimental verification can be done in future work. In addition, only two tilting state have been considered in the controller design and verification, turning with a constant speed and speed changes in a turn. The future research will also focus on other tilting states that are more complicated and challenged, such as maintaining the stability of NTV at highsider to avoid following down.

5. Conclusions

This chapter has reviewed the wheel and vehicle models of NTV and designed a nonlinear tilting controller for DTC-based mechanisms and two torque vectoring based drive assistance systems for torque control to help the rider in balancing the NTV during a turn. The nonlinear tilting controller has the capability to compensate the nonlinearities of vehicle tilting dynamics without the accurate vehicle model and been validated in simulation by comparing with the linear and gain-scheduling control approaches. The nonlinear tilting controller reduced the nonlinear tracking performance at different vehicle velocities and improved the yaw rate tracking performance as indicated in the results. The designed SATV and TCTV torque controllers are validated with the same vehicle model and compared with the traditional TV approach. The proposed controllers eliminated the counter-steering and improved the roll stability in balancing the vehicle. The peak tracking error, IAE and oscillation rate have been reduced by the tilting compensator and the vehicle dynamics are easier to be stable. Thus, the designed drive assistance systems can help riders with different experiences to balance the vehicle when driving an NTV.

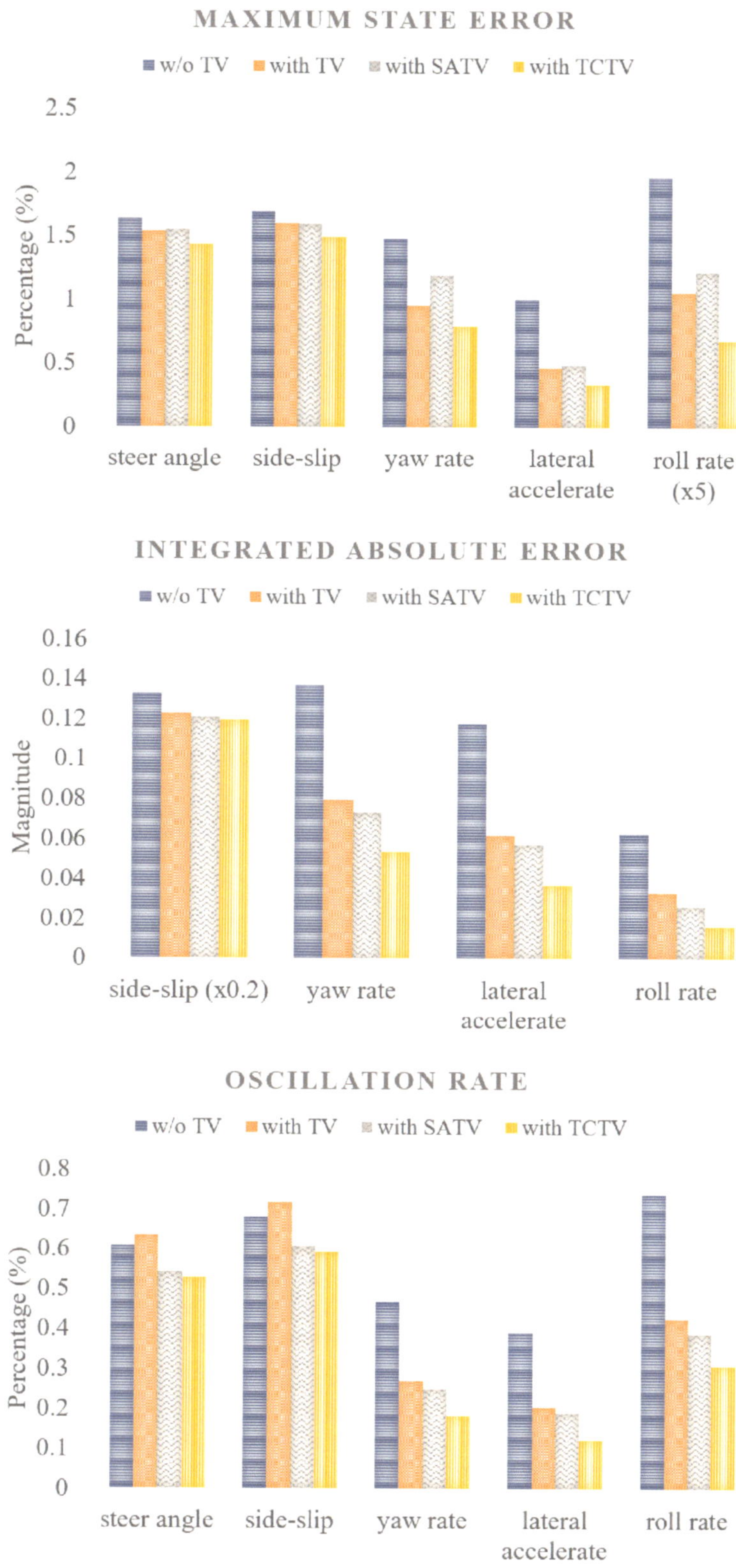

Figure 26.
The performance indices comparison among different torque controllers when accelerating in a turn.

Acknowledgements

The research presented in this chapter was undertaken as part of the Range of Electric Solutions for L Category Vehicles (RESOLVE) Project. Funded through the European Funding for Research and Innovation (Horizon 2020), Grant Number 653511.

Author details

Yaxing Ren
University of Warwick, Coventry, UK

*Address all correspondence to: yaxing.ren@warwick.ac.uk

References

[1] Gohl J, Rajamani R, Starr P, Alexander L. Development of a novel tilt-controlled narrow commuter vehicle. Minnesota, Minneapolis, MN, USA: Department of Mechanical Engineering University Reports CTS 06-05; 2006

[2] Kidane S, Alexander L, Rajamani R, Starr P, Donath M. Road bank angle considerations in modeling and tilt stability controller design for narrow commuter vehicles. In: American Control Conference. Minneapolis, MN, USA: IEEE; 2006. p. 6

[3] Van Poelgeest A. The dynamics and control of a three-wheeled tilting vehicle [PhD thesis]. University of Bath; 2011

[4] Robertson J. Active Control of Narrow Tilting Vehicle Dynamics [PhD thesis]. Bath, UK: University of Bath; 2014

[5] Berote JJH. Dynamics and Control of a Tilting Three Wheeled Vehicle. Ph.D. dissertation. Bath, UK: University of Bath; 2010

[6] Ren Y, Dinh Q, Marco J, Greenwood D, Hessar C. Nonlinearity compensation based tilting controller for electric narrow tilting vehicles. In: 2018 5th International Conference on Control, Decision and Information Technologies (CoDIT). Thessaloniki, Greece: IEEE; 2018. pp. 1085-1090

[7] Ren Y, Dinh Q, Marco J, Greenwood D. Torque vectoring-based drive: Assistance system for turning an electric narrow tilting vehicle. Proceedings of the Institution of Mechanical Engineers, Part I: Journal of Systems and Control Engineering. 2019;**233**(7):788-800. DOI: 10.1177/0959651818823589

[8] RESOLVE Project. 2018. Available from: http://www.resolve-project.eu/

[9] Fajans J. Steering in bicycles and motorcycles. American Journal of Physics. 2000;**68**(7):654-659

[10] Fabien C, Ph C, Lama M. Non-linear control of a narrow tilting vehicle. In: 2014 IEEE International Conference on Systems, Man and Cybernetics (SMC). San Diego, CA, USA: IEEE; 2014. pp. 2488-2494

[11] Pojani D, Stead D. Sustainable urban transport in the developing world: Beyond megacities. Sustainability. 2015; 7(6):7784-7805

[12] Li L, Lu Y, Wang R, Chen J. A three-dimensional dynamics control framework of vehicle lateral stability and rollover prevention via active braking with MPC. IEEE Transactions on Industrial Electronics. 2017;**64**(4):3389-3401

[13] Snell A. An active roll-moment control strategy for narrow tilting commuter vehicles. Vehicle System Dynamics. 1998;**29**(5):277-307

[14] Mourad L, Claveau F, Chevrel P. Direct and steering tilt robust control of narrow vehicles. IEEE Transactions on Intelligent Transportation Systems. 2014;**15**(3):1206-1215

[15] Van Den Brink CR, Kroonen HM. DVC[1]—The banking technology driving the CARVER vehicle class. In: 2004 7th International Symposium on Advanced. Vehicle Control. Arnhem, The Netherlands; 2004

[16] Kidane S, Alexander L, Rajamani R, Starr P, Donath M. A fundamental investigation of tilt control systems for narrow commuter vehicles. Vehicle System Dynamics. 2008;**46**(4):295-322

[17] So S-G, Karnopp D. Active dual mode tilt control for narrow ground vehicles. Vehicle System Dynamics. 1997;**27**(1):19-36

[18] Chiou J-C, Chen C-L. Modeling and verification of a diamond-shape narrow-tilting vehicle. IEEE/ASME Transactions on Mechatronics. 2008;**13**(6):678-691

[19] Furuichi H, Huang J, Fukuda T, Matsuno T. Switching dynamic modeling and driving stability analysis of three-wheeled narrow tilting vehicle. IEEE/ASME Transactions on Mechatronics. 2014;**19**(4):1309-1322

[20] Piyabongkarn D, Keviczky T, Rajamani R. Active direct tilt control for stability enhancement of a narrow commuter vehicle. International Journal of Automotive Technology. 2004;**5**(2): 77-88

[21] Mourad L, Claveau F, Chevrel P. Design of a two DOF gain scheduled frequency shaped LQ controller for narrow tilting vehicles. In: American Control Conference (ACC) IEEE. 2012. pp. 6739-6744

[22] De Novellis L, Sorniotti A, Gruber P. Wheel torque distribution criteria for electric vehicles with torque-vectoring differentials. IEEE Transactions on Vehicular Technology. 2014;**63**(4): 1593-1602

[23] Sawase K, Sano Y. Application of active yaw control to vehicle dynamics by utilizing driving/breaking force. JSAE Review. 1999;**20**(2):289-295

[24] Sawase K, Ushiroda Y, Miura T. Left-right torque vectoring technology as the core of super all wheel control (S-AWC). Mitsubishi Motors Technical Review. 2006;**18**:16-23

[25] Sawase K, Ushiroda Y. Improvement of vehicle dynamics by right-and-left torque vectoring system in various drivetrains. Mitsubishi Motors Technical Review. 2008;**20**:14

[26] Kang J, Heo H, et al. Control Allocation Based Optimal Torque Vectoring for 4WD Electric Vehicle. Technical Report, SAE Technical Paper. SAE International; 2012

[27] Yim S, Choi J, Yi K. Coordinated control of hybrid 4WD vehicles for enhanced maneuverability and lateral stability. IEEE Transactions on Vehicular Technology. 2012;**61**(4): 1946-1950

[28] Fallah S, Khajepour A, Fidan B, Chen S-K, Litkouhi B. Vehicle optimal torque vectoring using state-derivative feedback and linear matrix inequality. IEEE Transactions on Vehicular Technology. 2013;**62**(4):1540-1552

[29] Her H, Koh Y, Joa E, Yi K, Kim K. An integrated control of differential braking, front/rear traction, and active roll moment for limit handling performance. IEEE Transactions on Vehicular Technology. 2016;**65**(6): 4288-4300

[30] Koehler S, Viehl A, Bringmann O, Rosenstiel W. Energy-efficiency optimisation of torque vectoring control for battery electric vehicles. IEEE Intelligent Transportation Systems Magazine. 2017;**9**(3):59-74

[31] Hibbard R, Karnopp D. Twenty first century transportation system solutions-a new type of small, relatively tall and narrow active tilting commuter vehicle. Vehicle System Dynamics. Taylor & Francis; 1996; **25**(5):321-347

[32] So S-G, Karnopp D. Switching strategies for narrow ground vehicles with dual mode automatic tilt control. International Journal of Vehicle Design. 1997;**18**(5):518-532

[33] Rajamani R, Gohl J, Alexander L, Starr P. Dynamics of narrow tilting vehicles. Mathematical and Computer modelling of Dynamical Systems. 2003; **9**(2):209-231

[34] Kumar P, Merzouki R, Conrard B, Coelen V, Ould Bouamama B. Multilevel modeling of the traffic dynamic. IEEE Transactions on Intelligent Transportation Systems. 2014;**15**(3): 1066-1082

[35] Pacejka H. Tire and Vehicle Dynamics. 2nd ed. London, UK: Butterworth-Heinemann, Elsevier; 2005

[36] Svendenius J. Tire modeling and friction estimation [PhD theses]. 2007

[37] Ruggero F, Alessandro B. A virtual motorcycle driver for closed-loop simulation. IEEE Control Systems. 2006;**26**(5):62-77

[38] Kidane S, Rajamani R, Alexander L, Starr PJ, Donath M. Development and experimental evaluation of a tilt stability control system for narrow commuter vehicles. IEEE Transactions on Control Systems Technology. IEEE; 2010;**18**(6): 1266-1279

[39] Ohm DY. Analysis of pid and pdf compensators for motion control systems. In: Industry Applications Society Annual Meeting, Conference Record of the 1994 IEEE. Vol. 2. 1994. pp. 1923-1929

Chapter 8

The Impact of Virtual Environments for Future Electric Powered-Mobility Development Using Human-in-the-Loop: Part A - Fundamental Design and Modelling

Jun Jie Chong, Peter J. Kay and Wei-Chin Chang

Abstract

The use of virtual tools will be discussed across two complimentary chapters, Part A explores the fundamental concepts of electric vehicle systems modelling and a design procedure for human-in-the-loop virtual environments; Part B demonstrates how this architecture can be applied to assess energy optimization strategies. In Part A, this research investigates the design and implementation of simulation tools used to predict the energy consumption and strategic tool for the development of an electric vehicle. The case study used is an electric prototype race car for Ene-1 GP SUZUKA competition. Engineering effort is re-directed from physical product design, optimisation and validation to digital tools, processes and virtual testing. This virtual platform is characterised by the integration of two different simulation models—mathematical model of the electric vehicle systems represented by Matlab/Simulink, which accounts for the representation of the powertrain performance prediction that taking into account the resistance motion; and a virtual environment represented by Cruden Software, which accounts recreate topography of real world environment in a driving simulator and incorporate human driver behaviour.

Keywords: virtual environment, e-mobility, electric vehicle, virtual testing, Ene-1 GP

1. Introduction

Road transportation is one of the largest sources of greenhouse gases (GHG), responsible for climate change, and toxic emissions, responsible for poor human health. In many countries, the zero tailpipe emissions of electric vehicles (EVs) have gained attention from governments as emerging of state-of-the-art technology to reduce emissions from urban transport and improve air quality.

Morgan has carried out a research on global electric vehicle (EV) forecast (**Figure 1**) [1], the report indicates that the electric vehicles and hybrid electric

IntechOpen

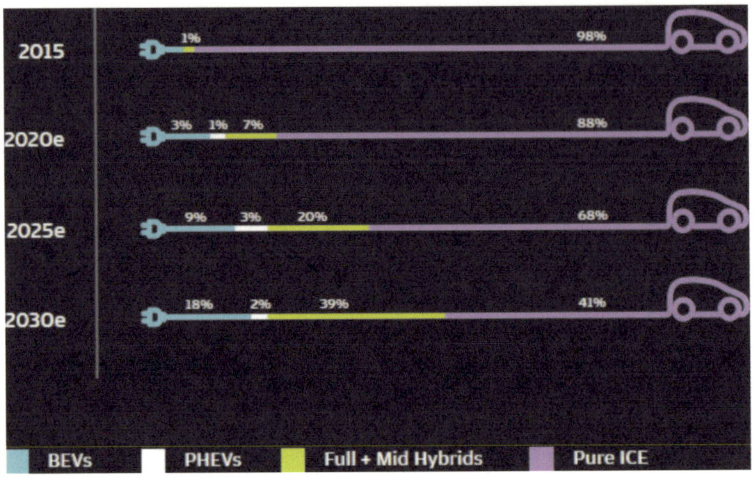

Figure 1.
Global electric vehicle forecast (source: Morgan estimates) [1].

vehicles (HEVs) is growing and by 2025, EVs and HEVs will account for an estimated 30% of all vehicle sales. It is also evident that, automotive manufacturers are preparing to phase out the internal combustion engine (ICE) vehicles and significantly investing in EVs to capitalise on their growth.

To assist this growth and facilitate the transition from conventional ICE vehicles towards full electrically driven vehicles, the performance (range, efficiency, and etc.) of EVs needs to be improved to mitigate the 'range anxiety' of future customers. In order to remain competitive, manufacturers are expected to streamline their new-product-development and manufacturing operations to trim down cost and introduce innovative products will be the most likely to enable them to meet the challenges. Therefore, automotive manufacturers are constantly advancing in new technologies to meet the four key requirements of the market demands, namely: time to market, price, quality, and variety.

For example, the model-based development methodology has been introduced to the vehicle development route in Ford product development system [2, 3] to enforce the replacement of hardware components by software models in the development process in order to reduce time and cost. Hardware-in-the-loop (HIL) is another important approach regularly implemented in automotive industry to serve as a comprehensive rapid prototyping and automated testing platform. HIL underpins the modelling technique that replaces a physical model, such as an electric vehicle drive train, with a mathematical representation that fully describes the important dynamics of the physical model. Poon et al. [4], highlighted several potential advantages of HIL tools could beneficial to industrial, enable:

- accelerated testing and validation process;
- reduced testing time needed in the lab;
- caused simulation of all operating points and scenarios that are difficult or impossible to recreate with a real system;
- fault injection capability;
- real-time access to all signals that are difficult to measure in a real system.

In addition to industrial applications, HIL also significantly contribute in the development of Advanced driver-assistance systems (ADAS). For example, A Robust Yaw Stability Controller was designed and tested for a road vehicle under a real time HIL testing [5]. The architecture of the HIL setup employed in that study included, the full-vehicle model ran on a dSpace DS 1103 simulator system, a controller was implemented on an xPC Target Box. The communication controller area network (CAN) bus communicate between the controller (xPC Targetbox) and the vehicle (DS 1103). The yaw rate sensor data, the steering input of the driver, and the vehicle speed can be acquired by the controller, and the steering actuation command can be sent to the vehicle dynamics simulator over CAN bus communication system is designed by which Robust Yaw controller performance is evaluated. Similar approach has been carried out by Gietelink et al. [6], developed an advanced driver assistant for cars and trucks using HIL simulation. In addition, the functioning of adaptive cruise control of automobile has been verified by HIL platform [7]. In Ref. [8], the authors developed a model predictive control approach for adaptive cruise control (ACC) systems and tested with HIL using relevant traffic scenarios, including Stop-&-Go.

According to Tang [9], virtual development is one of the technologies which will enable seamless integration of new processes and continuous improvements in automotive manufacturing. As shown in **Figure 2**, virtual environments (or called it as world models) are successfully integrated into the traditional V-shape model to support process development and validation of the design and performance of components and subsystems at each level. This concept has gained it attention over the recent years, Volvo have announced plans to reduce the development time of their vehicles from 42 to 20 months [10] from concept to start-of-production through the increased reliance on virtual tools, decoupled systems development, and common vehicle and powertrain architectures.

However, vehicle functions are becoming increasingly complexity, the main challenge for virtual development is on the integration of all individual sub-systems to recreate identical overall system behaviour in both virtual and under real world conditions. This chapter describes a generalised method for the fundamental design

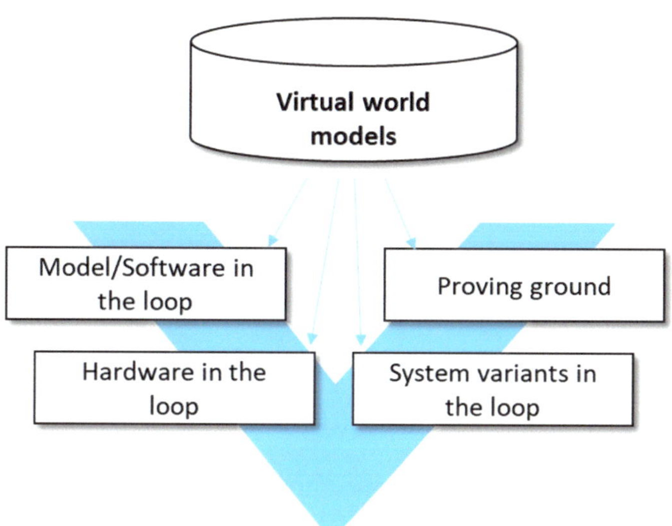

Figure 2.
Virtual world models embedded into V-cycle (adapted from AVL).

and modelling of vehicle system to be used as a plant model for Human-in-the-Loop application. Furthermore, the model was validated by with the real world data collected from the Ene-1 GP SUZUKA competition.

The paper is organised as follows. Section 2 presents the vehicle plant model development simulation. Simulation and results are given in Section 3. Finally, Section 4 summarises and concludes this chapter.

2. Fundamental design and modelling

This chapter focuses on the fundamental design and development of a vehicle model to be integrated as a digital tool in virtual environment for product design, optimisation and validation.

Several assumptions have been made to construct the models underlie this study:

- A simple vehicle longitudinal model in steady-state condition. This model assumes that the tyres are always contact with the road. Additionally, it is assumed that there is no vertical displacement and the effects of body roll or lateral load transfer is neglected.
- The electric machine is coupled with the inverter devoted as a traction system to convert energy from the vehicle's battery in order to power the drivetrain. For the sake of simplicity, the electrical dynamics of the system are not considered, so the requested torque is equal to the delivered torque, assuming that the torque is within the minimum and maximum limits. Efficiency maps are then used to determine the electric power consumed by the electric machine, given the rotational speed of the machine and the torque request.
- As battery thermal management is not implemented in the prototype, for the sake of simplicity, the thermal behavior of the battery is neglected.

Each sub-system of the vehicle plant model is discussed in the following sub-sections.

2.1 Design and development of vehicle model

The high-level strategy for modelling the car motion is as follows:

- Calculate the tractive force from the powertrain (Section 2.1.1)
- Calculate the sum of all the resistive forces acting on the vehicle (Aerodynamic drag, Rolling resistance, Force due to gradient) (Section 2.1.2)
- Calculate the net force acting on the vehicle (Section 2.1.3)
- Calculate the acceleration of the vehicle (Section 2.1.3)
- Determine the velocity and distance of the vehicle from the first and second integral of the acceleration respectively (Section 2.1.3)

To validate the model a non-commercial vehicle was selected. The reference vehicle chosen was an electric prototype race car for Ene-1 GP SUZUKA competition, as shown in **Figure 3**. The competition is a good case study as it encourages the

Figure 3.
Electric prototype race car for Ene-1 GP SUZUKA competition.

competitors to build an electric vehicle that is both fast and efficient. The vehicle was designed according to the rules of the competition.

2.1.1 Tractive force

Tractive force is the force produced by the electric machine applied between the tire and road surface to drive a car moving in the longitudinal direction and can be described mathematically as

$$F_{tractive} = \frac{T_t}{r_w} \tag{1}$$

where $F_{tractive}$ is the tractive force (N), T_t is the traction torque (Nm) from electric machine, r_w is the wheel radius (m).

2.1.2 Resistive forces

Figure 4 shows a generic vehicle travelling up an incline (with inclination angle θ) with the resistive forces acting on the vehicle. The primary aim of this vehicle model is to define the causality between vehicle speed to variations in machine output torque under the different environmental conditions, in which, the propulsion power (from electric machine) has to overcome the retardation forces from the environment to drive the vehicle forward.

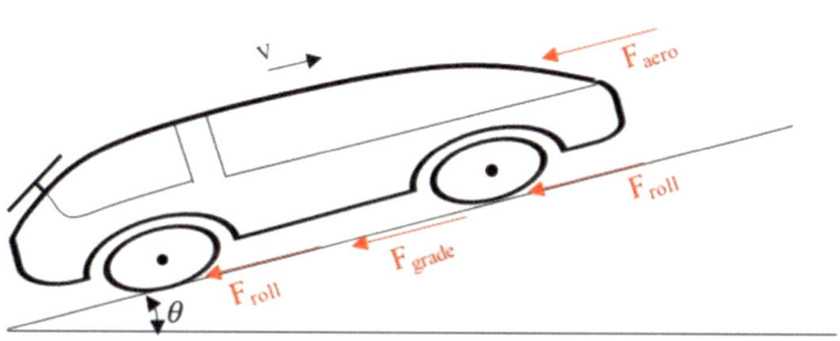

Figure 4.
Environment forces acting on the vehicle.

2.1.2.1 Aerodynamic drag force

The aerodynamics drag force mainly consists of two components; shape drag and skin friction. When a vehicle is travelling, there will be a high air pressure zone occurs in front of the vehicle and a low pressure zone appears behind the vehicle, these two zones will result in a resistive force resisting the car from moving forward. The resulting force on the vehicle is the shape drag. The second component of aerodynamic drag force named is as skin friction, is generated due to the fact that two air molecules with different speed create friction. This instance is important in the process of designing a prototype car for the Ene-1 GP SUZUKA competition because aerodynamic losses can account for up to 50% of total energy consumption [11], influence the operating range/speed and eventually affect the competition result. The aerodynamic drag force is defined as:

$$F_{aero} = \frac{1}{2}\rho C_d A (V_x)^2 \tag{2}$$

where ρ is density of air (kg/m^3), C_d is aerodynamic drag coefficient, and A (m^2) is the vehicle frontal area, and V_x is the speed of the vehicle in longitudinal direction.

Depending on the fidelity of the model this can be determined a number of ways. For approximations a value can be taken from **Table 1**, which shows the drag coefficient for a range of standard profiles.

No	Body	Status	Shape	C_D
1	Square rod	Sharp corner		2.2
		Round corner		1.2
2	Circular rod	Laminar flow		1.2
		Turbulent flow		0.3
3	Equilateral triangular rod	Sharp edge face		1.5
		Flat face		2
4	Rectangular rod	Sharp corner	L/D = 0.1	1.9
			L/D = 0.5	2.5
			L/D = 3	1.3
		Round front edge	L/D = 0.5	1.2
			L/D = 1	0.9
			L/D = 4	0.7
5	Elliptical rod	Laminar flow	L/D = 2	0.6
			L/D = 8	0.25
		Turbulent flow	L/D = 2	0.2
			L/D = 8	0.1
6	Symmetrical shell	Concave face		2.3
		Convex face		1.2
7	Semicircular rod	Concave face		1.2
		Flat face		1.7

Table 1.
Drag coefficients for standard profiles.

As it can been seen from the Eq. (2), the aerodynamic force is proportional to the vehicle speed. At constant speeds above 60 km/h, air resistance becomes the dominant retarding force for most vehicles [12]. This source of drag comes from both viscous skin friction as well as form of pressure drag. The form drag caused by the low pressure zone behind the vehicle is dominant factor contributing approximately 90 percent of the total drag [13]. In order to improve reduce the drag resistance, a CFD analysis (**Figure 5**) was performed to evaluate the aerodynamic characteristics of the vehicle and to change the line of its shape and to determine the drag coefficient more accurately.

2.1.2.2 Rolling resistance force

Hysteresis is the phenomenon in which the value of a physical property lags behind changes in the effect causing it. When the tire is rolling, rolling resistance is formed due to the deformation of the tires, Losses occur since the energy required to deform the part of the wheel in contact with the surface is more than that recovered when that part reverts to its original shape. The rolling resistance can be calculated in a simplified manner as a constant depending on the vehicle mass and the rolling resistance coefficient (C_{roll}) of the tyres.

$$F_{roll} = C_{roll} mg \tag{3}$$

where C_{roll} is a coefficient of rolling resistance, g the gravitational acceleration of 9.81 m/s^2 and m (kg) the mass of the vehicle. The rolling resistance coefficients given in **Table 2** are based on experimental results, many empirical formulae have been proposed for calculating the rolling resistance on a hard surface.

2.1.2.3 Gravitational force

As shown in **Figure 6**, the elevation changes act a major challenge for Ene-1 GP SUZUKA competition. The gravitational force is induced by gravity when driving

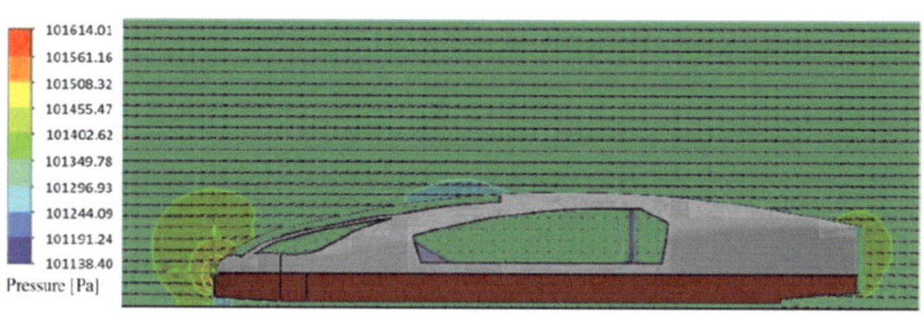

Figure 5.
Sample of the prototype CFD study.

Conditions	Rolling resistance coefficient
Car tires on concrete or asphalt	0.013
Car tires on rolled gravel	0.02
Tar macadam	0.025
Unpaved road	0.05
Field	0.1 - 0.35
Truck tires on concrete or asphalt	0.06 - 0.01
Wheels on rail	0.001 - 0.002

Table 2.
Rolling resistance coefficient at different conditions.

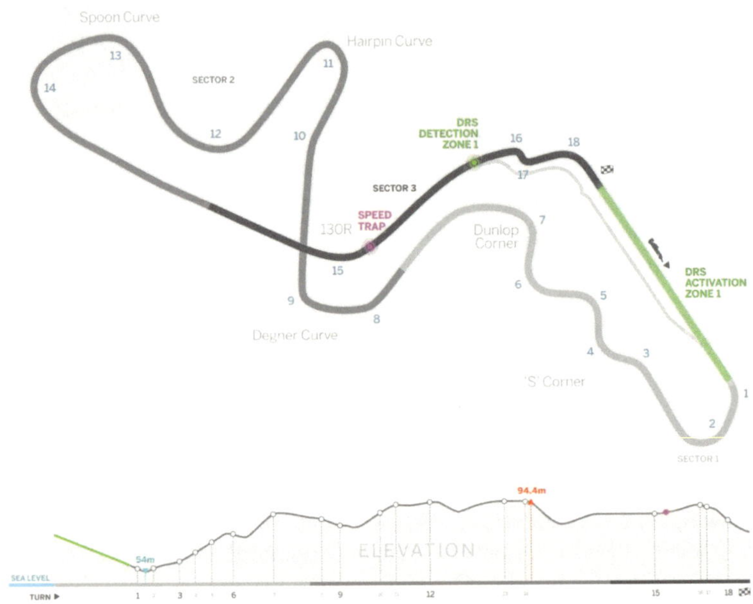

Figure 6.
SUZUKA F1 circuit with elevation profile [14].

on a non-horizontal road and depends on the slope of the road. The force is positive when the vehicle travels on an uphill section and negative on a downhill section. The gravitational force is given by:

$$F_{grade} = mg \sin \theta \tag{4}$$

where, g the gravitational acceleration of 9.81 m/s^2, m (kg) is the mass of the vehicle and θ (rad) is the inclination angle of the road.

2.1.3 Net force acting on the vehicle

From the Newton's Second Law of motion, the net force can be described mathematically as

$$F_{tractive} - F_{\text{Resistive}} = ma \tag{5}$$

where $F_{tractive}$ is the tractive force (N), $F_{\text{Resistive}}$is the total resistive forces (N), m is the mass of the vehicle (kg), a is the acceleration of vehicle (m/s^2).

2.2 Design and development of powertrain

The powertrain is a generic model for the components that transfer the torque from the electric machine to the wheels. In this section, it is simplified into an electric machine, transmission (sprockets and chain), and geared hub, all placed according to **Figure** 7. The output shaft of the electric machine is coupled to the sprocket and transmitted to the real wheel by a belt transmission. As shown in **Figure** 7, the internal geared hub is embedded into the real wheel, by using this setup, it will enable the driver has better shifting strategies. Each component has been modelled using a combination of mathematics and lookup tables to accurately capture the behaviour of the system. Specifications of the main components of this powertrain are given in **Table 3**.

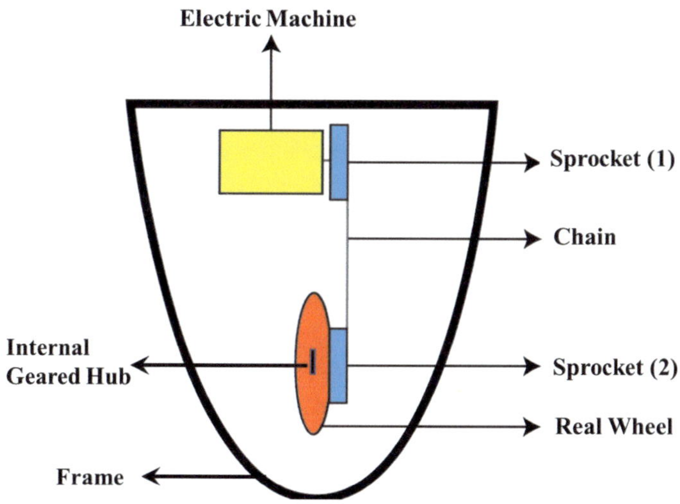

Figure 7.
Powertrain system layout.

Component	Specifications
Electric machine	S14502-500R (Tokushudenso)
Sprocket (1)	22
Chain	41.8 mm
Sprocket (2)	90
Internal geared hub	ALFINE SG-S501 (8-speed)

Table 3.
Powertrain system layout.

2.2.1 Electric machine

The electric machine model is built using the motor characteristic curve from TOKUSHUDENSO (Japan). The datasets contain sufficient data points to enable a 2D lookup table based performance models to be represented the internal operation of the components. The lookup table is implemented by using the driver demand (A) and electric machine speed (rpm) as inputs and to find the motor torque (N·m) output. A relative torque characteristic plot of the electric machine is shown **Figure 8**.

2.2.2 Driveline

The purpose of the transmissions submodel is to provide maximum vehicle performance and efficiency with wider gear ratios. Detailed ratio design is available in Naunheimer et al. [15] for gradient and speed requirements. Here, the minimum gear ratio for maximum grade climbing is used to evaluate lowest possible ratio, while top speed and driving torque are used to limit the top gear ratio.

$$Gear\ Ratio_{min} \geq \frac{(F_{resistive}) \times r}{n_s \times T_{max}} \tag{6}$$

where, $F_{resistive}(N)$ is the resistive forces applied on the vehicle, r (m) is the radius of the wheel, n_s is the efficiency of the transmission and $T_{\max}$ (Nm) maximum is the

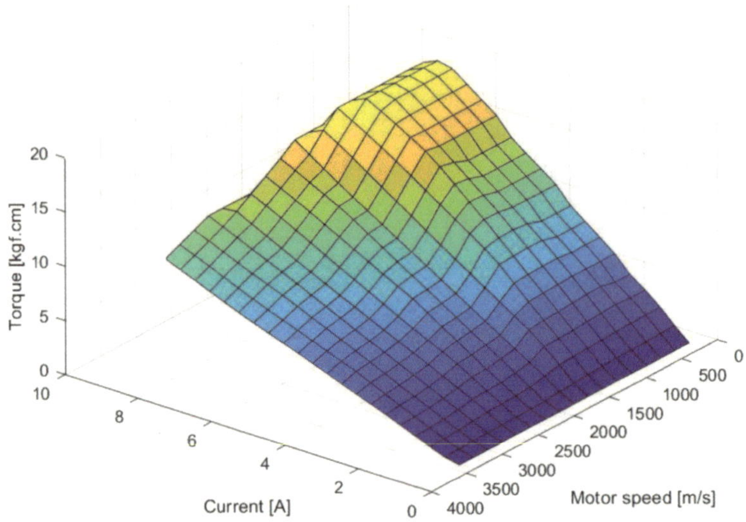

Figure 8.
2D lookup table for electric machine.

electric machine torque. The maximum gear ratio for top speed can be calculated by:

$$Gear\ Ratio_{max} \leq \frac{rpm \times \pi \times r}{30 v_x} \quad (7)$$

where, v_x (m/s) is the vehicle speed in longitudinal direction.

2.2.3 Battery model design and development

Energy storages are playing a significant role in today's electric vehicles (EVs) market penetration. Especially, in the realm of battery technologies, manufacturers are aiming to develop a high energy density, flexible and lightweight design for use in most handheld and portable electronics emergence of EVs. The high energy density of Lithium ion batteries (LIBs) have propelled them to become the main choice over the last decade. However, traditional electrode materials (such as commercial lithium cobalt oxide ($LiCoO_2$) cathodes and graphite anodes) have approached their theoretical capacities, limiting the energy density of LIBs (~260 W h kg^{-1}) (**Figure 9**) [18]. Based on the comparison of practical specific energies for several rechargeable batteries presented in **Figure 10** by Bruce et al. [16], Li-S and Li–O_2 are amongst the best of rough estimates, because so far there are few realistic prototypes on which to base such figures.

For future battery technologies, the issues of safety and reliability of the battery, accurate information about the state of charge (SOC) and its control still the main challenges. The primary function of the battery system model is to determine the resulting state of charge (SOC) of the battery system from the electrical current demanded from the electrical machine and the. The SOC of the battery is a key measure of performance, since it is directly related to the driving range of the vehicle. The academic literature presents several different battery models. The fidelity of these models differ in terms of the dynamics they represent and the parameterisation requirements they impose to support execution of the simulation. Given the data available to support battery simulation, a simplified stead-state

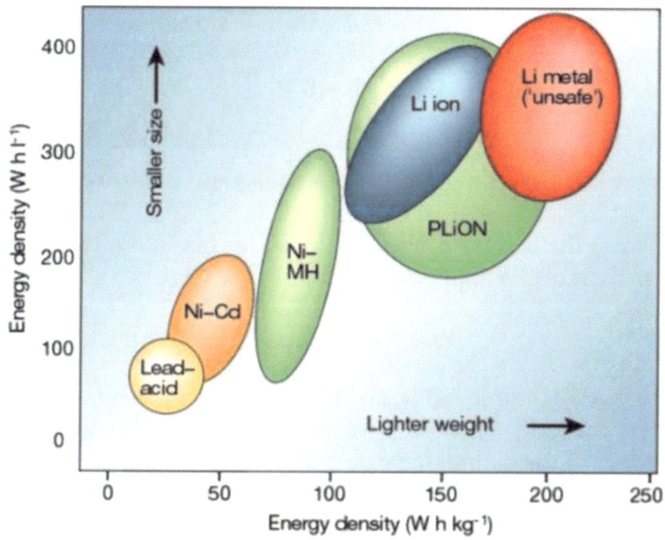

Figure 9.
Ragone plot of several of the battery technologies used in EVs [18].

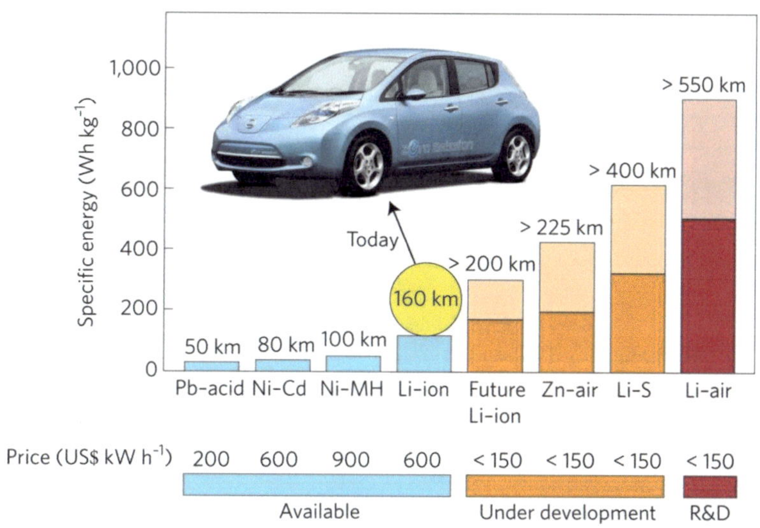

Figure 10.
Practical specific energies for some rechargeable batteries, along with estimated driving distances and pack prices [16].

model of the battery has been defined in the form of open circuit voltage (OCV) and cell impendence, both as a function of SOC.

In this competition, under the rules and regulations, the car is powered by Panasonic's AA size EVOLTA nickel metal hydride rechargeable batteries. They are connected 10 in series, forming four pairs. All four pairs are then coupled in series to form a 48 V battery pack. The specifications of the battery can be seen in **Table 4** and additional explanations of the battery setup are described in detail in the competition regulations [17].

To find the OCV dependence of the SOC, experimental was carried at constant current discharge and the result can be seen in **Figure 11**. The state-of-charge and depth-of-discharge depend on the integral of the current drawn or delivered to the battery, and can be calculated by:

$$DoD_{batt} = DoD_{ini} + \int \frac{i_{batt}}{Q_{batt}} dt \tag{8}$$

$$SOC = 1 - DoD_{batt} \tag{9}$$

where DoD_{batt} is depth-of-discharge, DoD_{ini} is the initial depth-of-discharge, SOC is the battery state-of-charge, i_{batt} is the equivalent battery current, and Q_{batt} is the equivalent battery capacity.

Figure 12 shows the battery model employed in this study, the battery terminal voltage, as a function of the applied current is:

$$V_{batt} = V_{oc} - I_{batt}R_{int} \tag{10}$$

The sign convention employed is: positive current represents current flowing out of the battery, negative current represents current flowing into the battery. Within the context of the complete powertrain model, the value of battery current is directly related to the torque generated by the electrical machine (either through

Type : Nickel-Metal Hydride Battery			Size : AAA Consumer Type
Capacity 1)	Typical		800mAh
	Minimum		750mAh
Nominal Voltage			1.2V
Charging Current x Time		Fast Charge 2)	800 mA x about 1.1h
Ambient Temp.	Charge Condition	Fast Charge 2)	0°C - 40°C
	Discharge Condition		0°C - 50°C
	Storage Condition	Less than 90days	-20°C - 40°C
		Less than 1year	-20°C - 30°C
Internal Impedance 3) (after discharge to E.V.=1.0V)			Approx. 40mΩ (at 1000Hz)
Weight 4)			Approx. 13 g
Size 4) :(Diameter) x (Height)			10.5(D) x 44.5(H) mm

1)Single cell capacity under the following condition.
Charge : 80mA×16h, Discharge : 160mA(E.V.=1.0V) at 20°C
2)Use recommended charging system.
3)After a few charge and discharge cycles under the above 1) condition.
4)With tube.

Table 4.
Battery Specifications.

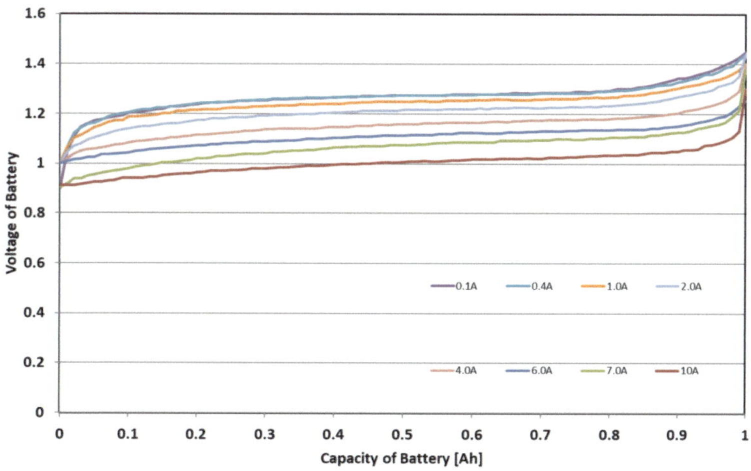

Figure 11.
Terminal voltage dependence of different constant discharge currents.

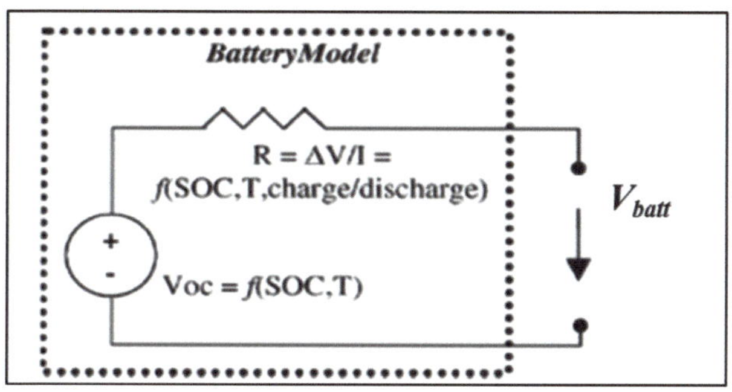

Figure 12.
Battery model employed in this study.

vehicle acceleration or regenerative braking) and any auxiliary load requested by the driver (e.g., vehicle lights etc.).

To calculate the battery SOC, the amount of current used during each simulation step is calculated and then subtracted from the initial state of charge at the start of the simulation and is calculated as:

$$SOC = \frac{SOC_{ini} - \int I_{batt} dt}{Ah * 3600} \tag{11}$$

where SOC is the instantaneous state of charge (0–1, 0 is empty and 1 is full), SOC_{ini} is the initial state of charge (0–1, 0 is empty and 1 is full), I_{batt} is the charge and discharge current (A) and Ah is the capacity of the battery (Ampere-hour).

3. Simulation and results

Figure 13 shows a summary of how the model is integrated. The target velocity is compared against the actual velocity with the fundamental PID controller. This the demands a current. This current is used to calculate both the torque from the powertrain and the SOC and voltage of the battery. The environmental forces, such as drag, are calculated from the vehicle velocity and position. The environmental forces and the force from the powertrain are used to calculate the vehicle

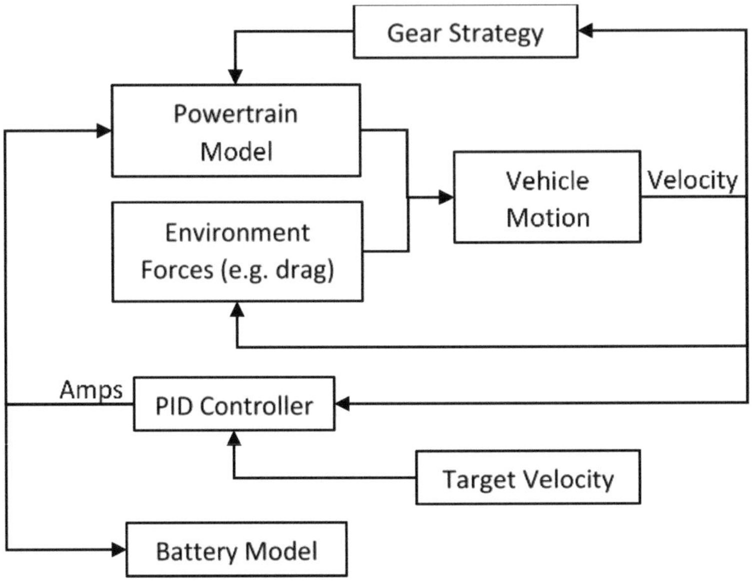

Figure 13.
Summary of model integration.

acceleration, velocity and position. The vehicle (motor) speed is used to determine the gear position from the gear strategy.

The PID controller is based on the proportional, integral and differential gains calculation that can be expressed on Eq. (12), where u(t) is controlled variable, e(t) is error value, *Kp* is proportional gain, *Ki* is integral gain and *Kd* is a derivative gain. Then, the PID controller parameters are stated in **Table 5**.

$$u(t) = K_p e(t) + K_i \int_0^t e(\tau) d\tau + K_d \frac{de(t)}{dt} \tag{12}$$

To validate the model simulations were conducted and compared to data collected from the real world from SUZUKA F1 Circuit. **Table 6** presents a representative data-set that can be employed to parameterise the vehicle model and facilitate initial concept simulation studies.

Figure 14 shows the variation of current with distance travelled for real-world data and simulation. The figure shows that there is broadly a very good agreement between the real-world data and the simulation.

Figure 15 shows the variation of vehicle velocity with distance travelled for real-world data and simulation. Broadly speaking there is good agreement between the real-world data and the simulation. However, there are some instances where there is some deviation between the real-world data and the simulation, in particular around 1500 m. It is thought that the discrepancy for this is due to the gear shift strategy.

Kp	**1**
Ki	**1**
Kd	**0**

Table 5.
Fundamental PID controller parameters.

Parameters	Symbols	Units	Values
Vehicle mass (with driver)	m	kg	77.48
Tyre rolling radius (front, rear)	R_{wheel_f}, R_{wheel_r}	m	0.17
Aerodynamic drag coefficient	C_d	—	0.62
Vehicle frontal area	A_f	m^2	0.5
Density of air	ρ	kg/m^3	1.25
Acceleration constant	g	kg/m^2	9.81
Tyre rolling resistance coefficient	C_{roll}	—	0.026
Front track width	tf	m	0.656
Wheelbase	l	m	2.956

Table 6.
Input parameters for vehicle model.

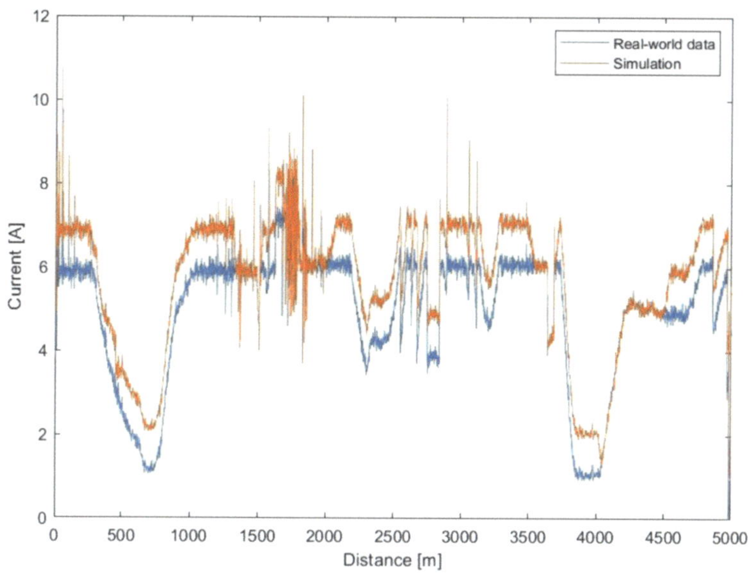

Figure 14.
Variation of current with distance travelled for real-world data and simulation.

For the real world data the gear that the vehicle was in is not known. Therefore, in the plant model a simple gear strategy algorithm was coded to change gear when the motor speed exceeded 3000 rpm or dropped below 1500 rpm.

However, in real life a human would not change gear based on such a simple strategy. A human would accept the motor speed to temporally increase above the desired threshold if the vehicle was, for example on a steep incline. This will be discussed in the next chapter.

Around 1500 m is where the steepest incline is on the Suzuka circuit. Therefore, it is likely that at this deviation in vehicle velocity is due to the fact that in the real-world the human driver is in an 'artificially' low gear to maintain the speed at the steepest part of the circuit.

Taking these factors into account there is reasonable absolute validation of the vehicle model to allow it to be used for future studies.

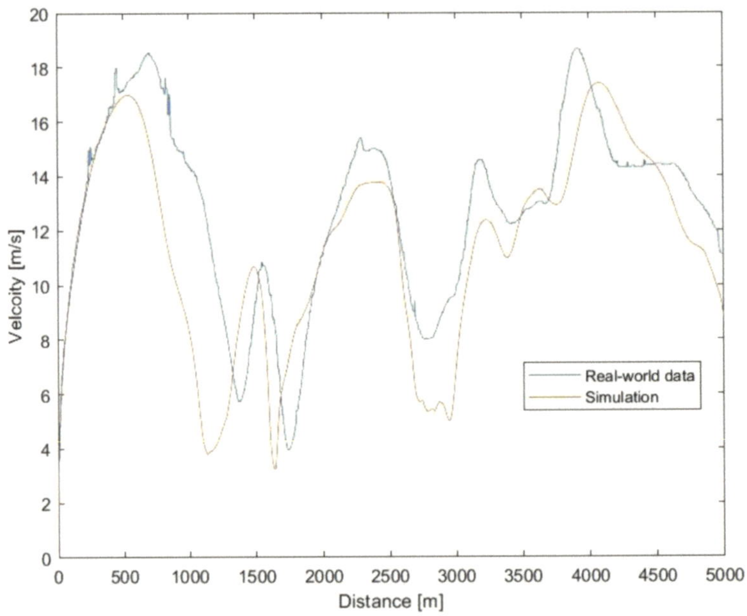

Figure 15.
Variation of vehicle velocity with distance travelled for real-world data and simulation.

4. Conclusion

This chapter presented the development of an electric vehicle model. A real-world non-commercial vehicle was used for validation. The main conclusions are:

- An electric vehicle model has been developed from first principles and commercially available data.
- The results show good absolute validation with real-world data
- The reason for areas where the validation is not as good, is due to the difference between the low fidelity gear strategy used in the model and the gear strategy used by the real-world driver.

Author details

Jun Jie Chong[1]*, Peter J. Kay[2] and Wei-Chin Chang[3]

1 Newcastle University in Singapore, Singapore

2 University of the West of England, Bristol, UK

3 Southern Taiwan University of Science and Technology, Taiwan

*Address all correspondence to: junjie.chong@newcastle.ac.uk

References

[1] Morgan JP. Driving into 2025: The Future of Electric Vehicles [Internet]. 2018. Available from: https://www.jpmorgan.com/global/research/electric-vehicles [Accessed: 03 December 2019]

[2] Liker JK, Morgan J. Lean product development as a system: A case study of body and stamping development at ford. Engineering Management Journal. 2011;**23**(1):16-28. DOI: 10.1080/10429247.2011.11431884

[3] Majumder A. Strategic metrics for product development at Ford Motor Company [Thesis]. Massachusetts: Massachusetts Institute of Technology; 2000

[4] Poon JJ, Kinsy MA, Pallo NA, Devadas S, Celanovic IL. Hardware-in-the-loop testing for electric vehicle drive applications. In: 2012 Twenty-Seventh Annual IEEE Applied Power Electronics Conference and Exposition (APEC); 5-9 February 2012; Orlando. IEEE; 2012. p. 2576-2582

[5] Guvenc BA, Guvenc L, Karaman S. Robust yaw stability controller design and hardware-in-the-loop testing for a road vehicle. IEEE Transactions on Vehicular Technology. 2008;**58**(2):555-571. DOI: 10.1109/TVT.2008.925312

[6] Gietelink O, Ploeg J, De Schutter B, Verhaegen M. Development of advanced driver assistance systems with vehicle hardware-in-the-loop simulations. Vehicle System Dynamics. 2006;**44**(7):569-590. DOI: 10.1080/00423110600563338

[7] Short M, Pont MJ. Assessment of high-integrity embedded automotive control systems using hardware in the loop simulation. Journal of Systems and Software. 2008;**81**(7):1163-1183. DOI: 10.1016/j.jss.2007.08.026

[8] Lee MH, Lee HM, Lee KS, Ha SK, Bae JI, Park JH, et al. Development of a hardware in the loop simulation system for electric power steering in vehicles. International Journal of Automotive Technology. 2011;**12**(5):733. DOI: 10.1007/s12239-011-0085-x

[9] Tang H. Manufacturing System and Process Development for Vehicle Assembly. Warrendale: SAE International; 2017

[10] SAE. Volvo's Rapid Strategy Aims at 20-Month Vehicle Development [Internet]; 2014. Available from: https://www.sae.org/news/2014/10/volvos-rapid-strategy-aims-at-20-month-vehicle-development [Accessed: 18 April 2019]

[11] Abo-Serie E, Oran E, Utcu O. Aerodynamics assessment using CFD for a low drag Shell eco-Marathon car. Journal of Thermal Engineering. 2017;**3**(6):1527-1536. DOI: 10.18186/journal-of-thermal-engineering.353657

[12] Schaltz E. Electrical Vehicle Design and Modeling. In: Soylu S, editor. Electric Vehicles: Modelling and Simulations. Rijeka: InTech; 2011. p. 1-24. DOI: 10.5772/958

[13] Sudin MN, Abdullah MA, Shamsuddin SA, Ramli FR, Tahir MM. Review of research on vehicles aerodynamic drag reduction methods. International Journal of Mechanical and Mechatronics Engineering. 2014;**14**(02):37-47. DOI: 10.26776/ijemm.04.01.2019.0

[14] Suzuka International Racing Course [Internet]. 2019. Available from: https://www.formula1.com/en/racing/2017/Japan.html [Accessed: 12 December 2018]

[15] Naunheimer H, Bertsche B, Ryborz J, Novak W. Automotive Transmissions: Fundamentals, Selection, Design and Application. 2nd ed. Heidelberg: Springer Science & Business Media; 2010. 717 p. DOI: 10.1007/978-3-642-16214-5

[16] Tarascon JM, Armand M. Issues and challenges facing rechargeable lithium batteries. Nature. 2001;**414**:359-367. DOI: 10.1038/35104644

[17] Bruce P, Freunberger S, Hardwick L, et al. Li–O_2 and Li–S batteries with high energy storage. Nature Materials. 2012;**11**:19-29. DOI: 10.1038/nmat3191

[18] Suzuka Circuit [Internet]. 2019. Available from: https://www.suzukacircuit.jp/ene1gp_s/ [Accessed: 10 February 2019]

Chapter 9

The Impact of Virtual Environments for Future Electric Powered-Mobility Development Using Human-in-the-Loop: Part B - Virtual Testing and Physical Validation

Jun Jie Chong, Peter J. Kay, Daisuke Yashiro and Wei-Chin Chang

Abstract

Electric vehicles are increasing in popularity worldwide, and there have been numerous advances in technology to increase the energy efficiency of the vehicle and reduce the range anxiety for the user. For example, the latest electric vehicle (Tesla model S, equipped by 100kWh battery) available in the market in 2019 is able to drive around 375 miles. However, human behavior such as driving strategy is an important issue that impacts on energy optimization and ultimately vehicle range. Human behavior is rather complex and is difficult to replicate with computer algorithms. Therefore, to fully assess the impact of a particular technology, the interactions between humans, vehicle, and the environment need to be examined simultaneously, through a Human-in-the-Loop approach. In this chapter, the results of investigating a human-in-the-loop test platform, which incorporate human-driving behavior and the vehicle characteristics, are presented. In addition, this chapter analyzes a driving strategy, using a Human-in-the-Loop approach, applied to optimizing the energy usage for an electric vehicle competition.

Keywords: human-in-the-loop, driving simulator, energy consumption prediction, energy management optimization

1. Introduction

Increasing levels of simulation is being adopted throughout the Automotive engineering industry. In a driving simulator, unlike the real world, the environment can be strictly controlled. For example, the weather conditions, traffic and topography can all be controlled. In other words, it makes it easier to decouple the driver behavior from the other variables. In this context, the use of a simulator would reduce the time and cost of optimizing the driver strategy and examine the performance of the vehicle.

Kemeny and Panerai [1] have defined a driving simulator as a system that provides a coherent multi-sensory environment for a driver to perceive and control virtual vehicle movements. The use of driving simulators is becoming ubiquitous within the automotive sector. Driving simulators offer significant advantages over real world testing such as controllability/reproducibility, cost and ease of data collection [2].

Coupled with the increased use of driving simulators, the importance of model-based development (MBD) [3, 4] has increased. The automotive industry has proactively adopted MBD for product development [5]. However, there are only few professional engineers involved in automobile development having sufficient experience for performing MBD. To cultivate our next generation, there are some competitions (e.g. Formula Students, Shell Eco Marathon, Ene-1 GP SUZUKA competition, etc.) designed for institutions that reinforce the MBD technique.

The automotive industry is currently undergoing a technical revolution, in the face of climate change, and is moving away from conventional ICE vehicles to full-electric vehicles.

However, there are well established challenges with transitioning to full electric vehicles, namely the speed/range trade-off due to the limited capacity of current battery technologies [6]. The higher the mean speed, the shorter the range and vice versa. The Ene-1 GP SUZUKA competition is a competition that encapsulates the challenges faced by this transition to full electric vehicles. The aim of the competition is to design and manufacture a full electric car that will complete three laps of the Suzuka F1 circuit in the quickest possible time. Therefore, much effort is spent by the competitors to determine the optimum driving strategy to minimize the limited electrical energy available. Development of these strategies will also be applicable to commercial electric vehicles.

As mentioned in the previous chapter, the virtual environment is gaining attention in automotive industry. The model is integrated into the V-cycle that underpins the model-based development technique for the development of electric vehicles, and it also contributes to reduce fuel consumption and greenhouse gas emissions by understanding driver behavior and driving style.

The behavior and style of the driver can have a significant impact on the overall efficiency of the vehicle. For example, Eco Driving has the potential to reduce emissions by 15% [7]. Several studies have examined driver behavior and how it impacts on vehicle efficiency. Most studies have focused on 'offline' simulations [8, 9] which used computer simulations to predict the optimum driver behavior.

Besides the physical fundamental modeling of each sub-system of the vehicle and its powertrain, carried out in the previous chapter, the driving strategy significantly impacts the range of battery electric vehicles (BEV). In motorsport environment, a commonly used parameter to identify the vehicle's performance is called 'minimum lap-time'. However, the traditional method to find the minimum lap-time is usually performed by a human racing-driver or using the predictive methods after collecting data from real world. However, this is an expensive way of testing, both in time and in money.

In this chapter, a Human-in-the-Loop approach will be used to examine the driving strategy for the Ene-1 GP SUZUKA competition. This challenge is similar to those related to commercial vehicles.

The literature related to human-in-the-loop is limited. Jameson et al. [10] used a human-in-the-loop to study the impact of different technologies to modify driver behavior of a conventional ICE vehicle. However, the research in using Human-in-the-loop to study the optimal driving behavior related to electric vehicles is limited. This research aims to address this.

The chapter is organized as follows. Section 2 provides the driving strategy review. Human-Hardware-in-the-Loop architecture given in Section 3. The limitations and further work are presented in Section 4. Finally, Section 5 concludes this chapter.

2. Driving strategy review

There have been numerous studies focused on improving driving strategy, so called 'eco-driving', with conventional ICE vehicles [11, 12]. However, the findings cannot be directly applied to hybrid electric vehicles (HEVs) or full-electric battery electric vehicles (BEVs). This is because the scope for eco-driving in conventional vehicles is fundamentally limited by the fuel-consumption which is principally affected by the throttle position and gear selection.

Hybrid vehicles require the most complex strategy to optimize energy usage, since the powertrain is the most complex, with the interplay between the ICE, electric motor and bidirectional energy flow from the batteries. Franke et al. [13] studied the driving strategies of efficient drivers of hybrid vehicles. They stated that key to efficient driving of hybrid vehicles is a technical knowledge of the powertrain system and being able to adapt the driving style to the changing environmental conditions. The work also presented recommendations that would allow drivers to adopt eco-driving behavior, and these included transparent and comprehensive feedback of the powertrain and energy usage.

With the regenerative braking function on BEVs, the studies examining eco-driving with internal combustion ICE vehicles cannot directly correlated to BEVs [14].

The BEVs powertrain architecture is less complicated when compared to other powertrains (ICEs, HEVs). However, as discussed, optimal driving strategies can significantly reduce the energy consumption and improve the range of BEVs. Therefore, in this section, a brief review of driving strategies is discussed.

Existing research of driving strategy investigations have mainly focused on safety concern, the most common commercially available today is Adaptive Cruise Control (ACC) which is based on robust control theory. In principle, ACC is able to control the vehicle by automatically adjusting the vehicle speed to maintain a safe distance from vehicles ahead only in the longitudinal direction. However, when the vehicle is experiencing a steering change, the target velocity of ACC needs to be lowered in order to keep the lateral acceleration within safety limits. Thus, unexpected change might cause ACC an error.

For this reason, Volkswagen introduced a Green Driving strategy [15] that integrated the predictive information from the upcoming diving environment into the ACC system. However, these systems only determine a local energy optimal driving strategy based on the restricted available information about the upcoming route without considering the impact of requirements of the global energy and time demand.

Zhang et al. [16], proposed a new approach to enhance the driver-individual driving strategy by observing and learning the driver's preferences. Two concepts of self-learning algorithms respectively based on dynamic programming (DP) and Q-learning (reinforcement learning) were developed to realize the approach.

In Choo's research [17], a machine learning software was developed at MIT for professional racing to improve the predictions of track position changes within a race. As shown in **Figure 1**, several factors that will possibly affect the racing performance were considered to analyze the system dynamics causality of race characteristics.

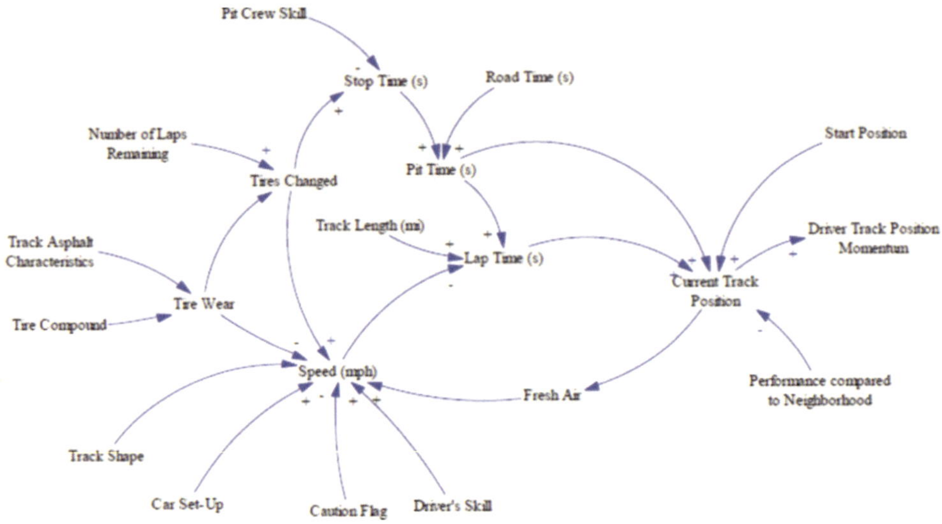

Figure 1.
System dynamics causal loop diagram of race characteristics affecting track position [17].

In general, the driver with the lowest cumulative total lap time throughout the course of a race is expected to win the race.

Hu et al. [18] investigated the driving control strategy of a battery electric bus by fuzzy logic control (FLC) algorithm, the findings show that proposed method has excellent output torque control of the electric machine and vehicle driving dynamics and acceleration performance are improved.

Another driving strategy example has been applied in Shell Eco-marathon competition [19] and successfully won the first place in the race. In their research, based on the experimental results, they examined the impact on racing time by varying the range of vehicle speeds in order to determine the most efficient fuel cell operating condition.

3. Human-hardware-in-the-loop architecture

Figure 2 shows the basic integration of the vehicle model (developed in the previous chapter) with a human driver. The input block reads the 'human' input data from the physical environment; these include steering angle, throttle and brake position and gear selection. The vehicle model then calculates the outputs; such as vehicle kinematics and powertrain properties. The vehicle model also includes information from the simulator domain, such as road gradient based on vehicle position. The outputs from the vehicle model are then used to determine the new vehicle position in the simulator domain. At the new position, information is passed back to the physical domain, such as the graphics for the user display, force feedback in the steering wheel, and some basic haptic feedback for the driver based on road noise.

For the experiments, in this chapter, simulation software called Panthera developed by Cruden was used. It runs the physics and graphics engine as well as communicating with the motion-based platform. It is used for real time simulations in Human-in-the-Loop simulators. **Figure 3** shows a screenshot of the virtual environment from inside the Cruden Software.

The physics model used was the same as discussed in the previous chapter. The vehicle was controlled using a Logitech G920 steering wheel, pedals and manual gear shifter. The steering wheel offered force feedback to improve the realism. The simulator did not use any motion or any other vestibular cues. Telemetry data from the simulation environment were recorded using the inbuilt application at a rate of 100 Hz.

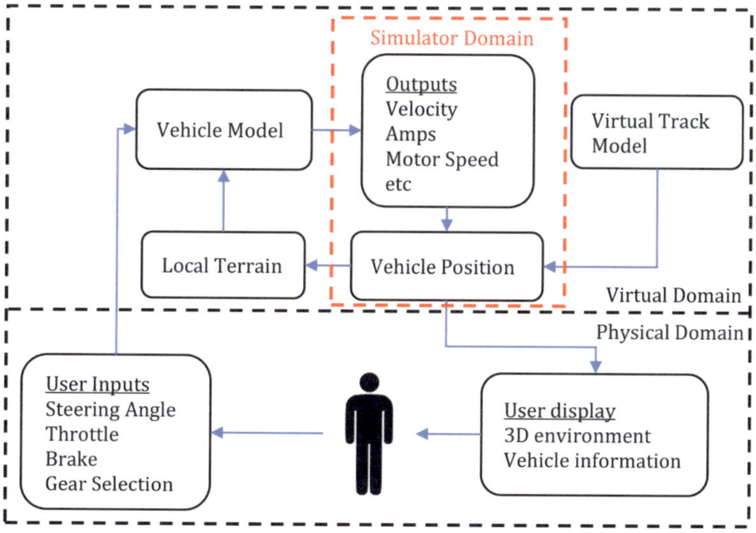

Figure 2.
Overview vehicle model integration in the simulator.

Figure 3.
(a) Simulator setup (b) screenshot of virtual environment of Suzuka F1 circuit.

Parameters	Symbols	Units	Values
Vehicle mass (with driver)	m	kg	77.48
Tyre rolling radius (front, rear)	R_{wheelf}, R_{wheelr}	m	0.17
Aerodynamic drag coefficient	C_d	—	0.62
Vehicle frontal area	A_f	m^2	0.5
Density of air	ρ	kg/m^3	1.25
Acceleration constant	g	kg/m^2	9.81
Tyre rolling resistance coefficient	C_{roll}	—	0.026
Front track width	t_f	m	0.656
Wheelbase	L	m	2.956

Table 1.
Input parameters for vehicle model.

The input parameters into the model are shown in **Table 1**.

The vehicle had eight gears with final ratios of 7.76, 6.35, 5.47, 4.81, 4.10, 3.34, 2.88 and 2.53.

In the first chapter, the battery current (throttle pedal) was inputted into the model based on real-world data. The gear was selected, based on simple thresholds of maximum and minimum motor speeds.

However, the advantage of having a human in the loop is that they can make more complex decisions such as

- Change gear based on a range of variables such as current speed, current throttle position.
- Use knowledge of the route (track) to predict/anticipate the vehicle inputs required to meet the coming features. E.g. lift off the throttle in anticipation of a down-hill slope.
- Drive to a prescribed strategy.

To demonstrate the applicability of human-in-the-loop virtual environments for Electric Powered-Mobility Development, a case study of an electric race vehicle driving around the Suzuka race track was used. For the current study the impact of different target currents on vehicle velocity and battery SOC were evaluated.

The human driver was free to decide on their gear strategy as long as the target battery current was adhered to. A simple algorithm in the plant model acted as a current controller and allowed the driver to vary the current from 0 to 10 A with the throttle pedal.

The target currents that the driver was instructed to maintain were 5A, 6A, 7A, and 8A, respectively.

4. Results and discussion

To assess the impact of including the human-in-the loop the results were compared against computer simulations in which the current was maintained constant at the specified value. In this analysis, the computer-in-the-loop will be referred to as the 'computer driver'.

Figure 4 shows the variation of current, vehicle velocity, gear position and battery state of charge against distance for both the human-in-the-loop and computer-in-the-loop fixed current runs for the 8A target.

Figure 4a Shows that the target current of 8A was mostly adhered to by the human driver, with notable exceptions such as the lifting off around 2.5 km on the approach to a hair-pin bend. The results show that although slightly different throttle and gear positions were selected the velocity profiles are very similar resulting in a similar trip duration. Consequently, overall the energy usage of the human driver is similar with the human driver achieving a slightly better final battery state of charge of 45% compared to 43%.

As discussed for the 8A target run, there was little impact of having the human-in-the-loop compared to having the compute-in-the-loop. However, due to the nature of the route used as a case study, Suzuka circuit, optimizing energy usage is difficult due to the topology, as this circuit has some challenging aspects to it. In particular, there is a steep climb of over 35 m starting at a distance of 900 m, as shown in **Figure 5**.

It was found that for both the computer-in-the-loop and the driver-in-the-loop tests that it was not possible to overcome the gradient and 35 m hill-climb for the 5A and 6A test runs. Therefore, at these sections it was required to increase the current (throttle position) to prevent the vehicle from coming to a standstill.

Figure 6 shows the variation of current, vehicle velocity, gear position and battery state of charge against distance for both the human-in-the-loop runs and computer-in-the-loop fixed current runs for the 5A target.

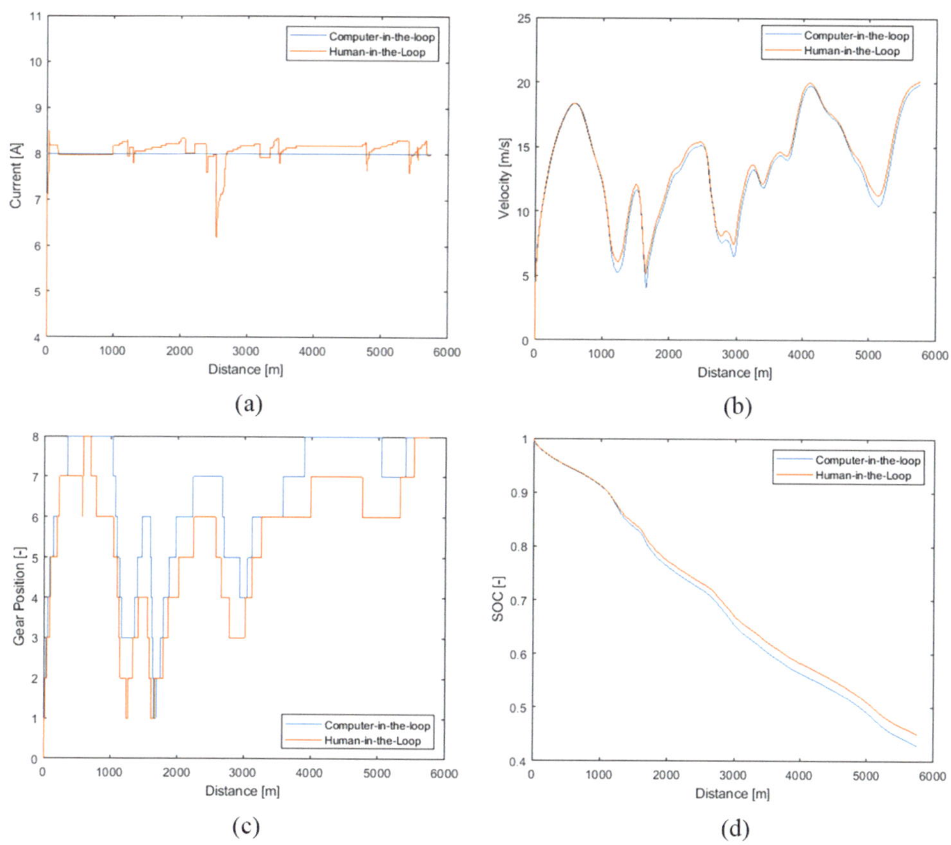

Figure 4.
Variation of (a) Current, (b) vehicle velocity, (c) gear position and (d) battery state of charge against distance for both the human-in-the-loop runs and fixed current runs for the 8A traget.

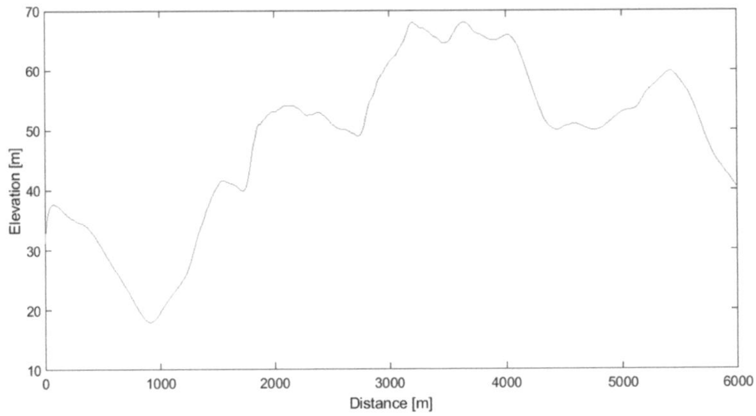

Figure 5.
Variation of elevation with distance.

Figure 6a shows that for the human driver a current of 10A (full throttle) was required to ensure that the vehicle made it over the crest of the hill. **Figure 4a** shows that full throttle was not necessary for the 8A test. This is due to the higher initial target current because of this the vehicle reaches the bottom of the climb with a higher velocity (**Figure 4b** and **6b**) and therefore has sufficient momentum to reach the top of the climb without demanding additional current.

The additional demand from the motor (full throttle) in the computer-in-the-loop tests was simulated with a simple step function. The duration of this step was determined by manually iterating the duration until the vehicle just cleared the crest of the hill. This was achieved as shown in **Figure 6b** where at around 1700 m the velocity of the vehicle almost reaches a standstill.

It can be shown in **Figure 6a** that full throttle is applied by the computer driver much earlier that the human driver. The reason for this is that the torque applied to the wheel of the vehicle is a function of the gear ratios as well as the torque from the motor.

Figure 6c shows that the human driver changed down through the gears much more quickly than the computer driver. The gear selection strategy for the computer-in-the-loop tests was the same as the one employed in the previous chapter. That is that the computer driver selected the gears to keep the motor speed within a set range. Because of this strategy, this meant that the computer driver changes gears later than the human driver.

This highlights the importance of having a human-in-the-loop whilst testing these technologies. The human driver had the foresight to select a lower gear in anticipation of the upcoming steep climb. To develop an algorithm for the computer-in-the-loop tests that could replicate this human behavior would be challenging. It would also unlikely to have the versatility to be applied to different scenarios where the circumstances are different.

The outcome of demanding full throttle for longer is that the final state of charge of the battery is much lower for the computer driver than for the human driver. **Figure 6d** shows that the final battery state of charge for the human and computer driver is 51% and 40% respectively.

Table 2 shows a summary of the final battery state of charge for each of the target currents for the human and computer driver. The results show that, generally, the state of charge increases as the target current decreases. This result is intuitive as running at a lower current will demand less energy from the battery.

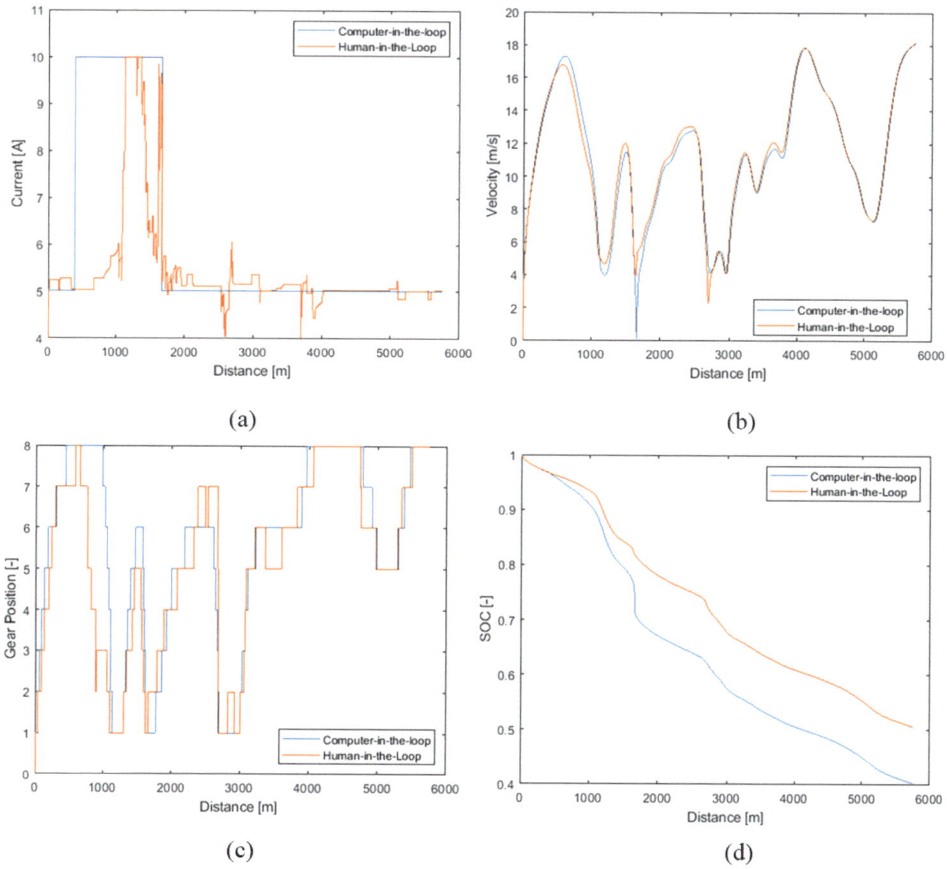

Figure 6.
Variation of (a) Current, (b) vehicle velocity, (c) gear position and (d) battery state of charge against distance for both the human-in-the-loop runs and fixed current runs for the 5A traget.

	Final SOC (%)	
Trial	**Computer driver**	**Human driver**
5A target	50.6	40.3
6A target	49.4	36.5
7A target	47.4	46.3
8A target	45.2	43.2

Table 2.
Summary of energy used for each of the target trials.

The results also show that in every case the human driver was able to optimize the energy better than the computer driver. The difference between human-in-the-loop and computer-in-the-loop is not constant. The impact of the human driver is more pronounced at lower target currents, a significant result when trying to optimize energy usage.

However, there is a tradeoff between target current and mean speed. In some instances, the mean velocity of the vehicle may not be a concern and therefore energy optimization is the primary concern. More likely, especially when this technology is being implemented in the public domain, a significant reduction in mean velocity is not acceptable, and a compromise is required.

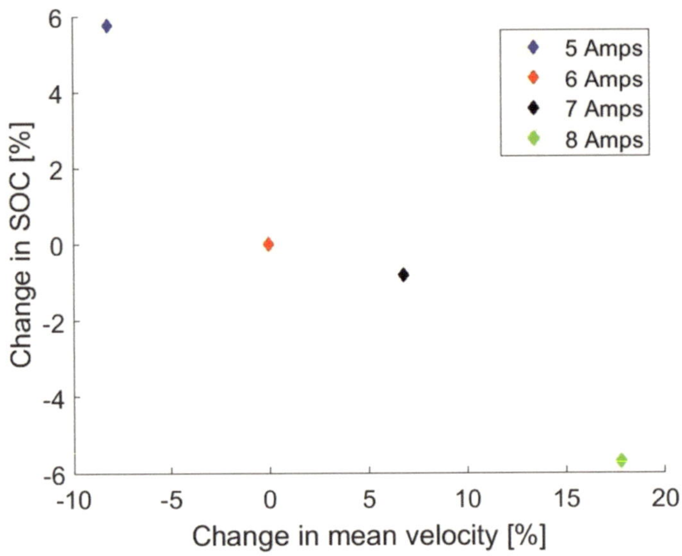

Figure 7.
Variation of change in final SOC with change in mean velocity for target currents of 5A, 6A, 7A & 8A.

Figure 7 shows the variation of change in final SOC with change in mean velocity for target currents of 5A, 6A, 7A & 8A. Generally, as expected, the final percentage change in final SOC decreases and the mean velocity increases as the target current is increased. However, there is a point of inflection in the trend between target currents of 6A and 7A. This shows, based on the data that increasing the target current from 6A to 7A has an appreciable change in mean velocity without a significant penalty in final SOC.

5. Limitations and further work

It should be noted that principally this study is a pilot study to examine the feasibility of using driving simulators to study the impact of having the human-in-the-loop to develop and optimize strategies to save energy from electric vehicles. This has been successful, and there are a number of limitations that prevent wider conclusions being drawn.

- The sample size is small with only four tests
- The sample were told about the research in advance and hence may have influenced their behaviour

To build on the research in this chapter a range of activities are planned as further work, these include:

- Testing alternative energy optimization strategies, such as limiting motor speed or vehicle acceleration.
- Include a larger range of tests, with statistical analysis, to limit the variability of human driver behavior.
- Incorporate hardware into the loop so that human-hardware-in-the-loop studies can be informed.

6. Conclusion

This chapter presents the evaluation of a Human-Hardware-in-the-loop architecture for Future Electric Powered-Mobility Development. To demonstrate the architecture a simple strategy of driving to a target battery current was employed. The results were compared to the results from computer-in-the-loop trials with the same conditions. The main conclusions are:

- The architecture allowed for good quality comparative data to be collected and driving strategies to be evaluated.

- The data showed that at relatively high target currents there was a marginal positive impact of the human driver compared to the computer driver.

- When running at low current targets, to conserve as much energy as possible, the human driver had a significant positive impact compared to the computer driver.

- The reason that the human driver was able to conserve more energy was due to anticipation of the upcoming gradient and better gear selection strategy.

- The data showed that, as expected, the mean driver velocity increased with the target battery current.

- The variation in mean velocity and SOC show an inflection point meaning that that increasing the target current from 6A to 7A has an appreciable change in mean velocity without a significant penalty in final SOC.

Author details

Jun Jie Chong[1,2], Peter J. Kay[3]*, Daisuke Yashiro[4] and Wei-Chin Chang[5]

1 Newcastle University, Singapore

2 Newcastle University, UK

3 University of the West of England, Bristol, UK

4 Southern Taiwan University of Science and Technology, Tainan, Taiwan

5 Mie University, Mie, Japan

*Address all correspondence to: peter2.kay@uwe.ac.uk

References

[1] Kemeny A, Panerai F. Evaluating perception in driving simulation experiments. Trends in Cognitive Sciences. 2003;**7**(1):31-37

[2] de Winter JCF, van Leeuwen PM, Happee R. Advantages and disadvantages of driving simulators: A discussion. In: Proceedings of Measuring Behavior 2012, 8th International Conference on Methods and Techniques in Behavioral Research. 2012

[3] Franceschini G, Macchietto S. Model-based design of experiments for parameter precision: State of the art. Chemical Engineering Science. 2008;**63**(19):4846-4872

[4] Sangiovanni-Vincentelli A, Di Natale M. Embedded system design for automotive applications. Computer. 2007;**40**(10):42-51

[5] Wan J, Canedo A, Al Faruque MA. Functional model-based design methodology for automotive cyber-physical systems. IEEE Systems Journal. 2015;**11**(4):2028-2039

[6] Schacht H, Kirchberger R, Winkler F, Schmidt S. Concept study of range extender applications in electric scooters. SAE Technical Paper. 2011-32-0592; 2011

[7] Liu H, Chen X, Wang X. Overview and prospects on distributed drive electric vehicles and its energy saving strategy. Przegląd Elektrotechniczny. 2012;**88**(7a):122-125

[8] Rios-Torres J, Sauras-Perez P, Alfaro R, Taiber J, et al. Eco-driving system for energy efficient driving of an electric bus. SAE International Journal of Passenger Cars-Electronic and Electrical Systems. 2015;**8**(1):79-89. DOI: 10.4271/2015-01-0158

[9] Kamal M, Mukai M, Murata J, Kawabe T. Influence of eco-driving on fuel consumption of vehicles in the following traffic. SAE Technical Paper 2011-08-0374; 2011

[10] Jamson SL, Hibberd DL, Jamson AH. Drivers' ability to learn eco-driving skills; effects on fuel efficient and safe driving behaviour. Transportation Research Part C: Emerging Technologies. 2015;**58**(D):657-668. ISSN: 0968-090X

[11] af Wåhlberg AE. Long-term effects of training in economical driving: Fuel consumption, accidents, driver acceleration behavior and technical feedback. International Journal of Industrial Ergonomics. 2007;**37**(4):333-343

[12] Van Mierlo J, Maggetto G, Van de Burgwal E, Gense R. Driving style and traffic measures-influence on vehicle emissions and fuel consumption. Proceedings of the Institution of Mechanical Engineers, Part D: Journal of Automobile Engineering. 2004;**218**(1):43-50

[13] Franke T, Georg Arend M, McIlroy R, Stanton N. Ecodriving in hybrid electric vehicles – Exploring challenges for user-energy interaction. Applied Ergonomics. 2016;**55**:33-45

[14] Kuriyama M, Yamamoto S, Miyatake M. Theoretical study on eco-driving technique for an electric vehicle with dynamic programming. In: 2010 International Conference on Electrical Machines and Systems; 10 October 2010; IEEE. 2010. pp. 2026-2030

[15] Knauder B, Karner M, Schratter M. Predictive longitudinal vehicle control based on vehicle-to-infrastructure communication. In: 2014 International Conference on Connected Vehicles and Expo (ICCVE); 3 November 2014; IEEE. 2014. pp. 258-263

[16] Zhang M, Storm K-F, Rausch A. Enhancement of driving strategy of electric vehicles by consideration of individual driver intention. In: 2017 EVS30 Symposium; 9-11 October 2017; Stuttgart, Germany

[17] Choo CL. Real-time decision making in motorsports: Analytics for improving professional car race strategy [doctoral dissertation]. Massachusetts Institute of Technology. 2015

[18] Hu J, Wu J, Peng H, Huang Q, Luo D, Fan L, et al., editors. Torque control strategy for battery electric bus. In: 2015 International Symposium on Material, Energy and Environment Engineering; 28 November 2015; Atlantis Press. 2015

[19] Wei Z, Song Y, Wang L, Zhang C, Cheng Lam JC, Teng X, et al. Design and energy efficiency analysis of a pure fuel cell vehicle for Shell eco racer. International Journal of Energy Research. 2019